This text covers the fundamentals of cryptography, which is concerned with methods of security in the storage and transportation of information.

Computers are now found in every layer of society, and information is being communicated and processed automatically on a large scale. Examples include medical and financial files, automatic banking, videophones, pay-tv, facsimiles, tele-shopping and global computer networks. In all these cases there is a growing need for the protection of information to safeguard economic interests, to prevent fraud and to ensure privacy. In this book, the fundamentals of secure storage and transportation of information are described. Among other things, attention is given to: symmetric (DES) and asymmetric (RSA) cryptographic algorithms, which are suitable for data security; methods for authentication; data integrity and digital signatures; key management; and network aspects.

The book will be of value to advanced students and researchers involved in data protection and information processing, including electrical engineers, mathematicians, business managers, system designers, application programmers, information analysts and security officers.

Basic Methods of Cryptography

Basic Methods of Cryptography

Jan C. A. VAN DER LUBBE

Faculty of Information Technology and Systems,
Delft University of Technology

Translated by Steve Gee

CAMBRIDGE
UNIVERSITY PRESS

PUBLISHED BY THE PRESS SYNDICATE OF THE UNIVERSITY OF CAMBRIDGE
The Pitt Building, Trumpington Street, Cambridge CB2 1RP, United Kingdom

CAMBRIDGE UNIVERSITY PRESS
The Edinburgh Building, Cambridge CB2 2RU, UK http://www.cup.cam.ac.uk
40 West 20th Street, New York, NY 10011–4211, USA http://www.cup.org
10 Stamford Road, Oakleigh, Melbourne 3166, Australia

Originally published in Dutch as *Basismethoden Cryptografie* by VSSD
and © VSSD 1994
First published in English by Cambridge University Press 1998 *as Basic
Methods of Cryptography*

Printed in the United Kingdom at the University Press, Cambridge

Typeset in Times

A catalogue record for this book is available from the British Library

ISBN 0 521 55480 2 hardback
ISBN 0 521 55559 0 paperback

Contents

Preface

As a result of current technological developments, the computer can now be found in all layers of our society and the possibilities for communication have grown immensely. At present, information is being communicated and processed automatically on a large scale. There are numerous examples: medical or fiscal computer files, automatic banking, video-phone, pay-tv, facsimiles, tele-shopping, global computer networks, etc. All these examples increasingly require measures for secure storage and transportation of the information. There are many reasons for this growing need. Protection of the information may be necessary to guard economic interests, to prevent fraud, to guarantee the privacy of the citizen, etc.

Cryptology is the science which is concerned with methods of providing secure storage and transportation of information in its widest sense.

In this book we will cover the fundamentals of secure storage and transportation of information, as they are currently being developed and used. The objective of this book is to allow the reader to become acquainted which the various possibilities of cryptology, and also with the impossibilities and necessary conditions involved in the use of cryptology.

This book is written for anyone who is in some way or other involved in protecting information processing and communication: engineers, system designers, application programmers, information analysts, security officers, EDP-auditors, etc.

This book has resulted from lectures given by the author to students of the Faculties of Electrical Engineering, Technical Mathematics and Informatics, Systems Engineering and Policy Analysis and Applied Physics of the Delft University of Technology and from the course in cryptology provided by TopTech Studies, which is responsible for the post-doctoral courses of the Delft University of Technology, and of which the author is the director.

The author wishes to thank dr.ir. J.H. Weber for his assistance during the lectures in cryptology at the Delft University of Technology and also all his TopTech cryptology course colleagues (in particular ir. R.E. Goudriaan of the International Nederlanden Bank), as they have taught the author a great deal about the practical aspects of the use of cryptology.

Delft   J.C.A. van der Lubbe
May 1997

Abstract

Chapter 1 focuses mainly on the role of cryptology within the total field of security. We will examine the various objectives of security and an initial summary of the available cryptographic methods is provided.

In Chapter 2 we will deal with the more classical forms of cipher systems, such as the transposition and the substitution ciphers. In addition, we will also take a look at the methods employed by cryptanalysts ('hackers') for cracking existing security measures.

In many cases, the strength of a cryptographic algorithm depends almost entirely on the obtainable level of security. However, since the term 'security' is itself far from clear, in Chapter 3 we will first deal with the concept of security, using terms from the field of information theory, and we will also pay attention to how security can be achieved.

One of the currently most popular cryptographic algorithms, which is based on enciphering with secret keys, is the DES algorithm. The principles of this algorithm are explained in Chapter 4.

Chapter 5 focuses on the use of shift registers for providing pseudorandom sequences, which can be used for generating keys as well as enciphering bit streams. In this chapter we will also study the term 'randomness'.

Chapter 6 is concerned with so-called public key systems; cryptographic algorithms with a secret and a public key. The RSA algorithm is an important example of such a system.

Chapter 7 deals with other types of cryptographic protection concerned with authentication and integrity. These items involve techniques which enable us to determine whether a transmitted message is intact and whether a message purported to be transmitted by some entity was really transmitted by that entity. Amongst other things we will examine digital signatures and zero knowledge techniques for identification.

In general ,we can say that no matter how good our cryptographic algorithms may be, the overall security always relies on the extent to which the secret keys remain secret. Chapter 8 therefore looks at the problem of key management, which is concerned with securely generating, distributing, etc., keys, as well as the specific aspects of the security of networks.

Finally, there are two appendices. Appendix A explains Shannon's measure of information and is meant for those who are not yet acquainted with the fundamentals of information theory. Appendix B covers several specific techniques for encrypting imagery.

Notation

A_d	Number of residues with d elements
C	Ciphertext
$C(\tau)$	Autocorrelation
CI	Coincidence index
CI$'$	Pure estimator of CI
d	Part of the secret key of RSA; number of elements of the residue
δ	Hamming distance
$DK(.)$	Decipherment with a symmetric algorithm using a key K
$dS_X(.)$	Decipherment with an asymmetric algorithm using a secret key S_X
D_L	Redundancy in a text of length L
e	Part of the public key of RSA
E	Expansion operation of DES
$E(.)$	Expectation
$EK(.)$	Encipherment with a symmetric algorithm using a key K
$eP_X(.)$	Encipherment with an asymmetric algorithm using a public key P_X
ε	Number of elements of an alphabet
ϕ	Euler totient function
$f(.)$	Characteristic polynomial
$f(.,.,.)$	Feedback function of a shift register
$F(.,.)$	Cipher function DES
$G(.)$	Generating function
χ	Chi-test
$h(.)$	Hash-code
H	Hypothesis
$H(.)$	Marginal information measure
$H(K/C)$	Key equivocation
$H(M/C)$	Message equivocation
$H(K/M,C)$	Key appearance equivocation
I	Identification sequence
IP	Initial permutation

IV	Initial vector
J	Jacobi symbol
K	Secret key of a symmetric algorithm
kDES	First k bits of the result of an encipherment using DES
K_i	Subkey of DES
K_{ij}	Key for Diffie–Hellmann protocol
KS	Session key
l	Length of a run
L	Length of a message
LC	Linear complexity profile
m	Number of sections of a shift register
$mgK(.)$	Result of a MAC using key K
M	Plaintext, original message
MK	Master key
mod	Modulo addition
N	Length of (pseudo)random sequence
$oK(.)$	Result of a one-way function using key K
p	Period of a shift register sequence; prime number
P_X	Pubic key of X for an asymmetric algorithm
Pe	Error probability
PeD	Error probability distance
q	Prime number
r	Total number of runs in a binary sequence
R	Random sequence
S	Knapsack sum; security event
S_X	Secret key of X for an asymmetric algorithm
s_i	Element of a shift register sequence; secret number for zero-knowledge techniques
T	Period
T_K	Encipherment transformation
TK	Terminal key
UD	Unicity distance
var	Variance

1

Introduction to cryptology

1.1 Cryptography and cryptanalysis

The title of this book contains the word *cryptography*. Cryptography is an area within the field of *cryptology*. The name cryptology is a combination of the Greek *cruptos* (= hidden) and *logos* (= study, science). Therefore, the word cryptology literally implies the science of concealing. It comprises the development of methods for *encrypting* messages and signals, as well as methods for *decrypting* messages and signals. Thus, cryptology can be divided into two areas: *cryptography* and *cryptanalysis*.

Cryptography can be defined more specifically as the area within cryptology which is concerned with techniques based on a secret key for concealing or enciphering data. Only someone who has access to the key is capable of deciphering the encrypted information. In principle this is impossible for anyone else to do.

Cryptanalysis is the area within cryptology which is concerned with techniques for deciphering encrypted data without prior knowledge of which key has been used. This is more commonly known as 'hacking'.

It is evident that cryptography and cryptanalysis are very closely related. One is only able to design good (sturdy) cryptographic algorithms when sufficient knowledge of the methods and tools of the cryptanalysists is available. The person responsible for the implementation of this type of security measure must therefore obtain this knowledge and be aware of the methods of a potential intruder. Obviously, successful cryptanalysis requires at least a fundamental insight into cryptographic algorithms and methods.

This book will focus mainly on cryptography.

A first impression of what a cryptographic algorithm does is given by considering the following situation, which also offers the opportunity of introducing some notation. Suppose *A* (the transmitter) wishes to send an enciphered message, i.e. secret code, to *B* (the receiver). Often, the original text or *plaintext* is simply denoted by the letter *M* of message and the encrypted message, referred to as the *ciphertext*, by the letter *C*. A possible method is for *A* to use a secret key *K* for *encrypting* the message *M* to the ciphertext *C*, which can then be transmitted and decrypted by *B*, assuming that B also possesses the secret key *K*. This is illustrated in Figure 1.1. *EK* represents the encryption of the message with the aid of *K*; the *decryption* of the message is represented by *DK*. Hereafter we will use the following notation:

$C = EK(M)$

(i.e. original text *M* is encrypted to ciphertext *C* with the secret key *K*);

$M = DK(C)$

(i.e. ciphertext *C* is decrypted to the original text *M* with the secret key *K*).

An example of what occurs at the transmitter and receiver is given by Figure 1.2. It is up to the reader to find the correct key. This should prove not too great a problem for those who can use a word-processor .

Figure 1.1. Cipher system.

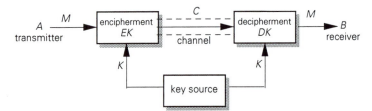

Figure 1.2. Example of encipherment and decipherment.

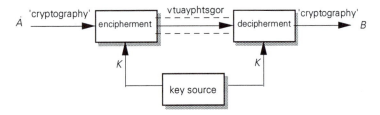

1.2 Aspects of security

Before proceeding with a description of the methods used in cryptography, we will first pay attention to the position and use of cryptography within the total concept of security. Here, three aspects play an important role, as illustrated in Figure 1.3.

The first question to be considered in practice is what purpose the security measures must serve. This unavoidably leads to some means of adequate threat analysis, which should provide a clear picture of what must be protected against whom or what.

Subsequently, the available means of security must be considered. This involves answering questions such as: How? With which security measures?

The third important aspect shown in Figure 1.3 has been labelled dimension. By this, we mean whether the security measures are designed for the prevention of or the correction of the damage caused by a security breach. We will return to this later.

The division of Figure 1.3 into three aspects can be extended to lower levels. If, for instance, we consider the purpose of the security measures, we can draw up a list of numerous possibilities against which security measures must be taken. Several examples are given below:

(*a*) reading or tapping data;
(*b*) manipulating and modifying data;
(*c*) illegal use of (computer) networks;
(*d*) corrosion of data files;
(*e*) distortion of data transmission;
(*f*) disturbance of the operation of equipment or systems.

The main issue of item (*a*) is *secrecy* and *confidentiality*. Confidentiality has always played an important role in diplomatic and military matters. Often information must be stored or transferred from one place to another,

Figure 1.3. Aspects of security.

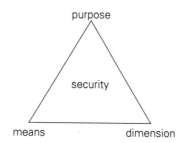

without being exposed to an opponent or enemy. Another example of how encipherment can be used to guarantee secrecy is in the communication between police patrols and the control room. Conversations are scrambled so that it is extremely difficult for outsiders to extract any relevant information from the transmitted messages. It is even conceivable that the simple fact of whether or not a message has been transmitted must also remain confidential. In this case, dummy messages can be transmitted in order to camouflage the real message.

So-called *key management* is also closely related to confidentiality. This area deals with generating, distributing and storing keys. It is obvious that no matter how strong a cryptographic algorithm may be, the final effectiveness of the algorithm depends largely on the obtainable secrecy level of the key. Once an intruder has managed to acquire a copy of the key, in principle, he is capable of decrypting the enciphered message. Therefore, key management must be considered as an essential element of the entire security plan.

Items (*b*)–(*d*) are primarily concerned with *reliability*. Take electronic banking, for instance. A bank requires some means of guaranteeing the authenticity of a financial transaction to prevent the wrongful withdrawl of large sums of money. Often the expression *integrity* is used as a measure of the genuineness of the data. Also, computer networks must be protected against intruders and unauthorised users. When one receives a fax message from a person *A*, one likes to be sure the fax was indeed written by *A* and that *A* is truly *A* and not an impostor. This is, in fact, an example of *authentication*, i.e. giving legal validity to the identity of the transmitter and determining the origin of the data. The above examples cover all aspects of security which are concerned primarily with reliability.

In items (*e*) and (*f*), a different aspect of the security of the information, its coninuity, is considered. Here, the data must be protected against deliberate disruption during its transmission and storage.

We can therefore distinguish between three different aims of security (see Figure 1.4). This also applies to the type of security used for a specific purpose (see Figure 1.5).

The phrases introduced here are self-evident. We can speak of *physical security* when a system is protected against the physical entry of an intruder, for instance by using metal containers, certain plastics or temperature or vibration sensors. However, in this book we will deal with *hardware* and *software based security*; i.e. cryptographic algorithms and methods.

No matter how high the standard of physical and hardware and software security of a system, the safety of the information cannot be guaranteed without sufficient organisational measures. So-called *organisational secur-ity* ensures that conditions are created which allow the physical and hardware and software security measures to be fully effective. If, in practice, certain security measures prove complicated or confusing to the user, naturally this will introduce the risks of neglect and carelessness. Therefore, one must always bear in mind that human beings will always be present somewhere in the chain, regardless of the level of automation of a security system.

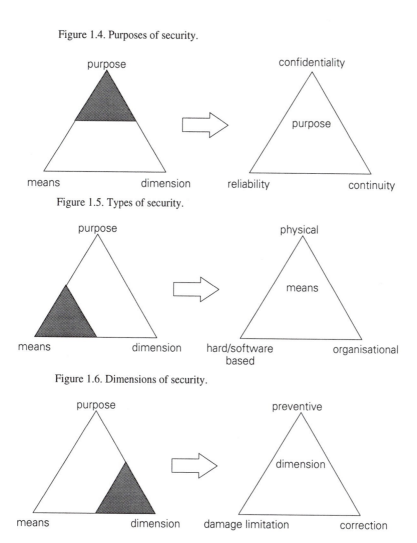

Figure 1.4. Purposes of security.

Figure 1.5. Types of security.

Figure 1.6. Dimensions of security.

The last aspect we will consider here is the dimension of the security. Again, we can divide this into three (see Figure 1.6). In the first instance, cryptography is applied as a *preventive* measure. We are attempting to minimise the chance of anything happening to the transmitted information. This can be accomplished by using strong cryptographic algorithms and protocols and installing adequate physical and organisational measures. However, by definition, *absolute security measures* do not exist. In practice, the chance of a mishap occurring can be minimised, but can never be reduced to zero. Therefore, another aspect of the dimension of security is *damage limitation*. If the chance of a mishap occurring is not zero, we can at least ensure that the resulting damage remains as limited as possible. For example, if someone manages to break into a computer file, it is possible to ensure that he gains access to only a small part of the entire file. Or if a key falls into the wrong hands, it must never be possible to decrypt all messages with that key, but only a fraction of them, etc. The final aspect of the dimension of security is *correction*. If something happens to the encrypted information, it must be possible to correct this quickly. For instance, there must always be some means available of rendering a key useless, just in case the key falls into the hands of an unauthorised person. Also, when vital information is damaged, it must be possible to reconstruct this information easily.

Obviously, in any practical situation a trade-off between the listed aspects of security must be made.

In addition, the economic facet of the security measures has to be taken into consideration. This is the relation between the desired level of security, the value of that which is being secured, or of that against which is being secured, and the investments necessary to obtain the desired level of security or to gain access to the information.

In the preceding text several applications of cryptography were mentioned. These can be divided into two groups, i.e. applications related to the storage of information and those related to the transportation of information.

Nowadays information is mostly stored in computer systems, on either disc or magnetic tape. The method of storage is often public knowledge and only the key is kept secret. As this type of data is usually stored for a considerable length of time, a cryptanalytic attack is attractive; the cryptanalyst can take his time finding the key. Consequently, this situation requires a relatively high level of security.

On the other hand, when the data are transmitted (e.g. TV, satellite), they are available to the cryptanalyst for only a short period of time and, in addition, the key can easily be changed regularly. Obviously, the cryptanalyst can record a transmitted message, but this does not necessarily help him to decrypt other transmitted messages if the key is frequently altered.

Moreover, communicated messages are often meaningful for only a short period of time, as the information ages or becomes obsolete (e.g. news, weather information, etc.). For this reason, such data communication usually requires a lower level of security and therefore also lower investment.

The costs of the security measures must also be viewed in a different light. Consider, for instance, cable television. The cable company will obviously profit from good security measures which prevent as much illegal viewing as possible. However, the price of decrypting the TV signal must still remain reasonable, for both the cable company and the consumer. Clearly, most consumers, who can be regarded as honest subscribers, will only pay a limited contribution towards security measures they did not ask for themselves. Furthermore, the level of security is sufficient if a potential viewer must make a larger investment than the ordinary subscription to be able to view the programmes illegally.

1.3 Cryptanalytic attacks

Let us consider a cryptographic algorithm which requires the use of a secret key. With regard to Figure 1.1, we can speak of a cryptanalytic attack when an intruder tries to discover the contents of a message or the secret key by other means than straightforward random attempts. Clearly, the intruder will find it more interesting if he can discover the key itself, rather than occasionally disclosing the plaintext, as then, hopefully, other ciphertexts can be decrypted as well. One method of finding the correct key is by simply trying all possibilities, until the correct key is found. This is called an *exhaustive key search*. However, this is not really a cryptanalytic attack in the true sense, as generally we expect a cryptanalyst to behave more 'intelligently'.

As far as real cryptanalytic attacks are concerned, we can distinguish between three types, depending on the level of information available to the cryptanalyst, see Figure 1.7. These three types are:

(*a*) an attack based solely on the ciphertext: (*ciphertext-only-attack*);

(*b*) an attack based on a given plaintext and the corresponding ciphertext: (*known-plaintext-attack*);

(*c*) an attack based on a chosen plaintext and corresponding ciphertext: (*chosen-plaintext-attack*).

When the attack is based on the ciphertext alone, the cryptanalyst only has access to the encrypted signal. With the use of the necessary statistics and by analysing apparent patterns in the signal the cryptanalyst must attempt to decipher the hidden message (plaintext) and more importantly, the key. This is often the case when analysing enciphered speech, tapping car telephones, etc. It is clear that this situation is the least favourable for a cryptanalyst.

A far more favourable situation can be found when the cryptanalyst can obtain information on the corresponding plaintext, in addtion to what he already knows about the ciphertext. If a relation can be found between a certain part of the ciphertext and the plaintext, then this knowledge may be used to decrypt other sections of the ciphertext, or even to find the key. In order to obtain information on both the ciphertext and the plaintext the cryptanalyst must gain access to (a part of) the cipher system or its users. Every financial transaction, for instance, contains information on the payer and the payee. If a cryptanalyst has inside information on how the information on the parties involved is enciphered in the message, he can attempt to decipher the remaining part of the ciphertext.

The most favourable situation is one in which the cryptanalyst can select a certain plaintext and generate the corresponding ciphertext. By choosing the correct plaintext and corresponding ciphertext, he can decipher parts of the text which are still encrypted, or even find the key. For example, a word processor which stores files in an encrypted form is an easy target for a chosen-plaintext-attack.

Ideally, a cryptographic system must be able to withstand all three types of attack, although in practice this is often difficult to realise. A system which appears capable of resisting ciphertext-only-attacks, may prove sensitive to know chosen-plaintext-attacks. However, a system which can withstand a chosen-plaintext-attack is usually regarded as of a higher standard than a system which can only stand an attack based on the ciphertext alone.

Figure 1.7. Cryptanalytic attack.

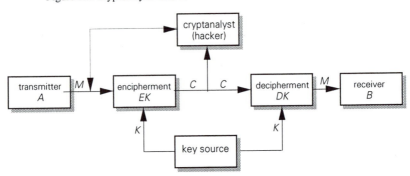

The above text focuses mainly on cryptanalytic attacks which form a breach of confidentiality. Attacks which affect the reliability (integrity and authenticity) will be described in Chapter 7.

2

Classical cipher systems

2.1 Introduction

In this chapter we will examine several classical cipher systems, which are often based on methods of encipherment with a very long history. Nowadays, though, most of these methods have become less popular; they were frequently used during the second world war, but since computers have become available to cryptanalysts, their applicability has diminished. However, this does not imply that a description of classical cipher systems is merely of historical interest. On the contrary, although the classical cipher systems are rarely used on their own, they are often incorporated in more modern crypto-systems, either in a cascaded form, or in combination with other methods.

We can distinguish two types of classical cipher system:
– transposition systems;
– substitution systems.

A transposition cipher is based on changing the sequence of the characters of the plaintext; the characters themselves remain unchanged. A substitution cipher does not alter the order of the characters of the plaintext, but replaces the original characters with others.

We will examine these two cipher systems in more detail in the following sections.

2.2 Transposition ciphers

In the previous section we mentioned that a transposition cipher only alters the order of the characters of the plaintext. This is performed in blocks of characters and is demonstrated by the following example.

plaintext: THE MEETING HAS BEEN POSTPONED UNTIL NEXT MONTH
divided into
blocks: THEME ETING HASBE ENPOS TPONE DUNTI LNEXT MONTH
ciphertext: MEETH NGIET BESHA OSPEN NEOTP TINDU XTELN THNMO

The plaintext of this example is divided into blocks of five letters. We can say that here, the period is equal to 5. Within each block the order of the letters is changed according to the key 4 5 3 1 2, so, with regard to the original block, the 4th and 5th characters have been exchanged with the 1st and 2nd and the 3rd remains in the same position.

Often a so-called key-word is used so that the key is easily remembered. Here, a suitable key-word is, for instance, 'stock'. The key is given by the alphabetical order of the letters of the key-word.

The encryption of a message with a transposition cipher can, in fact, be regarded as the transposition of the columns of a matrix, as is demonstrated by the following example. The blocks of letters are now placed below each other, rather than next to each other.

plaintext: THE MEETING HAS BEEN POSTPONED UNTIL NEXT MONTH
key-word: STOCK, KEY 45312

THEME	MEETH
ETING	NGIET
HASBE	BESHA
ENPOS	OSPEN
TPONE	NEOTP
DUNTI	TINDU
LNEXT	XTELN
MONTH	THNMO

ciphertext: MEETH NGIET BESHA OSPEN NEOTP TINDU XTELN THNMO

The ciphertext is obtained by changing the columns according to the alphabetical order of the letters of the key-word.

Consider a message M with a total length of $L = nT$ letters, in which T is the period and n a positive integer. If the message is divided into blocks of length T, it can be represented by:

$$M^{nT} = [a_1,a_2,\ldots,a_T][a_{T+1},\ldots,a_{2T}]\ldots[a_{(n-1)T+1},\ldots,a_{nT}],$$

or, by introducing matrix notation:

$$M^{nT} = \begin{bmatrix} a_1 & a_2 & \ldots & a_T \\ a_{T+1} & a_{T+2} & \ldots & a_{2T} \\ \vdots & & & \vdots \\ a_{(n-1)T+1} & \ldots & \ldots & a_{nT} \end{bmatrix}$$

or even as:

$$\underline{M}^T = [\underline{a}_1,\underline{a}_2,\ldots,\underline{a}_T],$$

in which for every $i = 1,\ldots,T$ the column $\underline{a}_i$ is given by:

$$\underline{a}_i = (a_i,a_{T+i},\ldots,a_{(n-1)T+i})^t.$$

The corresponding ciphertext C can be found by merely interchanging the columns of the matrix, or, in formal notation:

$$\underline{C}^T = [\underline{a}_{k(1)},\underline{a}_{k(2)},\ldots,\underline{a}_{k(T)}]$$

in which k is a permutation of $(1,\ldots,T)$.

We can now calculate the total number of possible keys, which is equal to $T!$ or, more precisely, $T! - 1$, because one key will always produce a ciphertext identical to the plaintext.

Obviously, the period T must be large. Our example has a period of 5, which means that there are $5! - 1 = 119$ possibilities for the key. This value is rather small and a cryptanalyst who knows the period T will find no difficulty in quickly decrypting the ciphertext.

On the other hand, if a large period is used to increase the number of different keys and make it more difficult for a cryptanalyst, the rightful receiver is forced to remember a very long key, which introduces new risks.

Cryptanalysts are faced with two problems. First, they must find the period T, which involves trying all possible combinations of n and T, which satisfy $L = nT$, with L equal to the length of the message. If dummy letters have been added to the original text, then more combinations of n and T must be considered. The second problem is to find the key in a structured manner, preferably without having to try all possible permutations first.

To solve these two problems, the cryptanalyst can benefit from the linguistic characteristics of the language in which the plaintext is written. Certain letters are used more frequently than others, as can be seen in Figure

2.1, in which the relative letter-frequencies of several languages are plotted. This kind of plot can also be made for the frequency of combinations of two letters. Regarding a text in this manner, we find that vowels tend to be surrounded by consonants and vice versa. This means that vowels are generally distributed evenly throughout a text. Therefore, when the ciphertext is put into a matrix form, a period T should be chosen which produces the most even distribution of the vowels of the text across the columns of the matrix. The text can now be deciphered more easily if the columns are exchanged in such a way that the most frequently occurring letter combinations appear first.

Figure 2.1. Relative frequency of occurrence of letters for several languages.

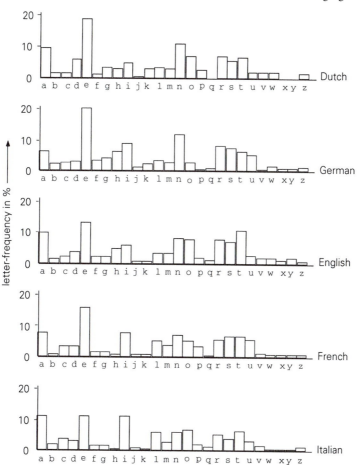

2.3 Substitution ciphers

A substitution cipher is based on replacing the characters of the plaintext with other characters. Assuming the plaintext is based on an alphabet of 26 letters, a substitution cipher can be described by the following:

alphabet of the plaintext: $A = [a_1,...,a_{26}]$
alphabet of the ciphertext: $B = [b_1,...,b_{26}]$
plaintext: a_3, a_{23}, a_9, a_{17}, a_4
ciphertext: b_3, b_{23}, b_9, b_{17}, b_4

The most straightforward substitution cipher is the Caesar substitution, named after the Roman Emperor Julius Caesar (100–44 BC). The substitution alphabet is obtained by simply shifting the original alphabet a given number of characters, with respect to the original alphabet. In the following example the alphabet is shifted three places.

original alphabet A:

A B C D E F G H I J K L M N O P Q R S T U V W X Y Z

substitution alphabet B:

D E F G H I J K L M N O P Q R S T U V W X Y Z A B C

plaintext: PLEASE CONFIRM RECEIPT
ciphertext: SOHDVE FRQILUP UHFHLSW

If the characters of the plaintext alphabet and ciphertext alphabet are numbered and denoted by i and j respectively, then in the above example, for all $i = 1,...,26$: $j = i + 3 \pmod{26}$. Mod 26 implies that the left part and right part of the equation may only differ by a multiple of 26. In a more general form, $j = i + t \pmod{26}$, in which t represents the number of characters the two alphabets are shifted.

An important characteristic of the Caesar substitution is the fact that the order of the characters of the substitution alphabet remains unchanged. The total number of keys is no more than 26, so this cipher can very easily be cracked; once a single letter of the ciphertext can be related to a letter of the plaintext, the system breaks down. If the message is sufficiently large, it is all the more straightforward to find such a relation; simply note the most frequently occurring letter and the chances are that this is equal to the letter e of the original plaintext, assuming that this was written in English.

If the letters of the cipher alphabet are placed in random order, instead of simply being shifted with respect to the original alphabet, the number of keys increases to 26!. This makes decipherment considerably more difficult than in the case of Caesar substitution. For example:

original alphabet *A*:

    A B C D E F G H I J K L M N O P Q R S T U V W X Y Z

substitution alphabet *B*:

    E S T V F U Z G Y X B H K W C I R J A L M P D Q O N

plaintext: PLEASE CONFIRM RECEIPT
ciphertext: IHFEAF TCWUYJK JFTFYIL

It is even possible for the substitution alphabet to consist of entirely different symbols from the original alphabet, as for instance in Figure 2.2.

Clearly, the key of this example is far too complicated to be remembered easily and therefore key-sentences are often used. Each time a new letter appears in the key-sentence, it is added to the substitution alphabet . When all the different letters of the sentence have been recorded, the remaining letters of the alphabet are placed behind this list in their usual order. Consider the following example:

key-sentence: THE MESSAGE WAS TRANSMITTED AN HOUR AGO

original alphabet *A*:

    A B C D E F G H I J K L M N O P Q R S T U V W X Y Z

substitution alphabet *B*:

    T H E M S A G W R N I D O U B C F J K L P Q V X Y Z

plaintext: PLEASE CONFIRM RECEIPT
ciphertext: CDSTKS EBUARJO JSESRCL

Despite the 26! possibilities, finding the correct key will still be relatively easy, since languages generally contain a high level of redundancy. Also, the commonest letters, such as e, t, n, r, o, a, etc., can always be found with comparatively little effort by considering the letter-frequency distribution.

We can conclude that, in general, substitution methods as described above are not very resistant to attacks, since the characteristics of the language can still be extracted from the ciphertext. This can be avoided by applying more than one substitution cipher. This procedure is referred to as a *polyalphabetical substitution*, as opposed to the *monoalphabetical substitution* of the examples above. A well-known example of a polyalphabetical substitution is the Vigenère system, which was devised in France in 1568 by Blaise de Vigenère. This system uses a different Caesar substitution for each letter. For example, the first letter is shifted by 10 positions, the second by 17, etc.

Encryption based on the Vigenère system is often performed with the aid of a so-called *Vigenère table* (see Table 2.1) and a key-word. The top row of the Vigenère table consists of the letters of the plaintext alphabet and the first column contains the letters of the key-word. A text can be enciphered using the following procedure.

The key-word is repeated below the plaintext as in the example below. A letter of the ciphertext is equal to the letter located at the intersection of the column designated by the letter of the plaintext and the row designated by the letter of the key.

Figure 2.2. Excerpt of a cipher as used for the communication between the Dutch Viceroy Willem Lodewijk and his commander Fredrich von Vernou.

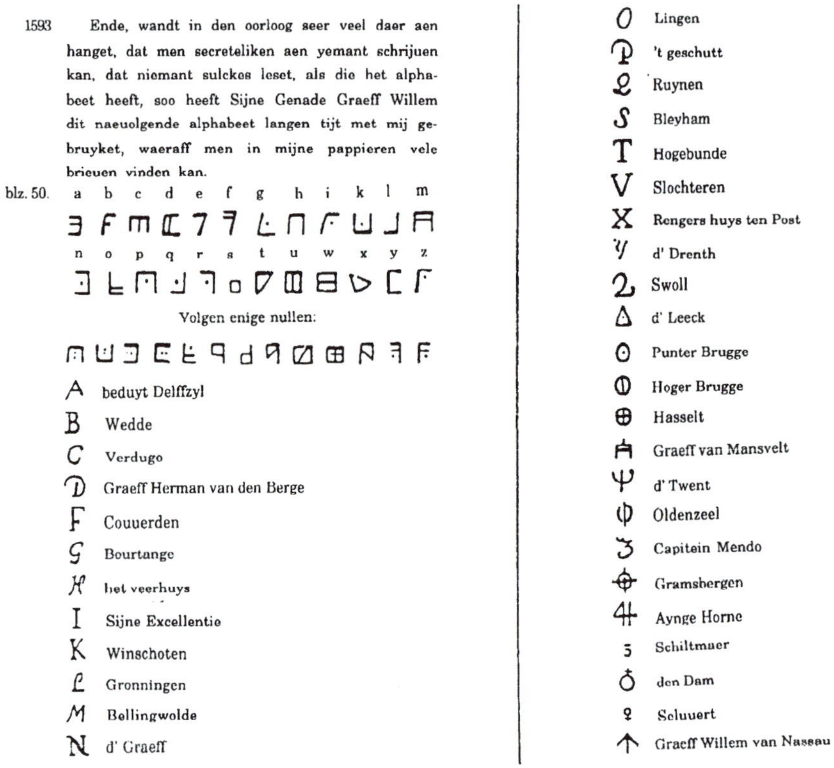
1593 Ende, wandt in den oorloog seer veel daer aen hanget, dat men secreteliken aen yemant schrijuen kan, dat niemant sulckes leset, als die het alphabeet heeft, soo heeft Sijne Genade Graeff Willem dit naeuolgende alphabeet langen tijt met mij gebruyket, waeraff men in mijne pappieren vele brieuen vinden kan.

blz. 50. a b c d e f g h i k l m

n o p q r s t u w x y z

Volgen enige nullen:

A beduyt Delffzyl
B Wedde
C Verdugo
D Graeff Herman van den Berge
F Couuerden
G Bourtange
H het veerhuys
I Sijne Excellentie
K Winschoten
L Gronningen
M Bellingwolde
N d' Graeff
O Lingen
P 't geschutt
Q Ruynen
S Bleyham
T Hogebunde
V Slochteren
X Rengers huys ten Post
Y d' Drenth
Z Swoll
d' Leeck
Punter Brugge
Hoger Brugge
Hasselt
Graeff van Mansvelt
d' Twent
Oldenzeel
Capitein Mendo
Gramsbergen
Aynge Horne
Schiltmaer
den Dam
Seluuert
Graeff Willem van Nassau

plaintext: PLEASE CONFIRM RECEIPT
key: CRYPTO CRYPTOC RYPTOCR
ciphertext: RCCPLS EFLUBFO HCRXWRK

Clearly, a given letter of the plaintext is represented by different letters in the ciphertext, depending on the letters of the key-word, thus concealing linguistic characteristics more effectively than any of the previous methods.

The number of monoalphabetical substitutions on which the Vigenère system is based is equal to the length of the key word. Here, the number of monoalphabetical substitutions is five and consequently, five rows of the table have been used. Obviously, if a cryptanalyst can discover the length of the key-word, this knowledge will be of great help in finding a solution to the cryptogram.

Someone using this system will generally attempt to employ as many different rows of the table as possible. One way of ensuring this is to use the plaintext itself, in addition to the key-word. This is demonstrated by the following example. The key is constructed by placing the letters of the plaintext itself after the key-word CRYPTO. The major problem of transposition and substitution ciphers is successfully to hide the statistical

Table 2.1. The Vigenère table.

```
A B C D E F G H I J K L M N O P Q R S T U V W X Y Z
B C D E F G H I J K L M N O P Q R S T U V W X Y Z A
C D E F G H I J K L M N O P Q R S T U V W X Y Z A B
D E F G H I J K L M N O P Q R S T U V W X Y Z A B C
E F G H I J K L M N O P Q R S T U V W X Y Z A B C D
F G H I J K L M N O P Q R S T U V W X Y Z A B C D E
G H I J K L M N O P Q R S T U V W X Y Z A B C D E F
H I J K L M N O P Q R S T U V W X Y Z A B C D E F G
I J K L M N O P Q R S T U V W X Y Z A B C D E F G H
J K L M N O P Q R S T U V W X Y Z A B C D E F G H I
K L M N O P Q R S T U V W X Y Z A B C D E F G H I J
L M N O P Q R S T U V W X Y Z A B C D E F G H I J K
M N O P Q R S T U V W X Y Z A B C D E F G H I J K L
N O P Q R S T U V W X Y Z A B C D E F G H I J K L M
O P Q R S T U V W X Y Z A B C D E F G H I J K L M N
P Q R S T U V W X Y Z A B C D E F G H I J K L M N O
Q R S T U V W X Y Z A B C D E F G H I J K L M N O P
R S T U V W X Y Z A B C D E F G H I J K L M N O P Q
S T U V W X Y Z A B C D E F G H I J K L M N O P Q R
T U V W X Y Z A B C D E F G H I J K L M N O P Q R S
U V W X Y Z A B C D E F G H I J K L M N O P Q R S T
V W X Y Z A B C D E F G H I J K L M N O P Q R S T U
W X Y Z A B C D E F G H I J K L M N O P Q R S T U V
X Y Z A B C D E F G H I J K L M N O P Q R S T U V W
Y Z A B C D E F G H I J K L M N O P Q R S T U V W X
Z A B C D E F G H I J K L M N O P Q R S T U V W X Y
```

parameters of the text. One solution is to ensure that the characters of the ciphertext have a uniform distribution. The characters of the plaintext can be coded in a certain manner, e.g. according to the Huffmann coding, to obtain a uniform distribution. This distribution is preserved when a transposition or substitution cipher is applied to the plaintext.

2.4 The Hagelin machine

In the 1930s, the Swede Boris Hagelin invented a machine which is capable of generating enciphered text based on polyalphabetical substitutions (see Figure 2.3). This machine is called the Hagelin cryptograph (or, M-209 machine) and was used by the American army until approximately 1950.

The cryptograph employs a polyalphabetical substitution method which relies on the so-called *square of Beaufort* (see Table 2.2), which is comparable to the Vigenère table. The substitution is given by:

$$j = t + 1 - i \ (\mathrm{mod}\ 26),$$

in which i is the alphabetical position of the plaintext character, t the row of the Beaufort table and j the alphabetical position of the encrypted symbol.

Figure 2.3. The Hagelin machine. The lower photograph clearly shows the coding wheels (1) and drum (2) (photographs by Facilitair Bedrijf TU Delft, Photographic Service).

Example

Assume the plaintext is equal to the sequence of letters 'SECRET' and that the following values of t are used: 2, 15, 8, 7, 3 and 1. We then find:

$$
\begin{aligned}
&\text{S:} && i = 19, && t = 2 && \Rightarrow j = 10 \Rightarrow \text{J} \\
&\text{E:} && i = 5, && t = 15 && \Rightarrow j = 11 \Rightarrow \text{K} \\
&\text{C:} && i = 3, && t = 8 && \Rightarrow j = 6 \Rightarrow \text{F} \\
&\text{R:} && i = 18, && t = 7 && \Rightarrow j = 16 \Rightarrow \text{P} \\
&\text{E:} && i = 5, && t = 3 && \Rightarrow j = 25 \Rightarrow \text{Y} \\
&\text{T:} && i = 20, && t = 1 && \Rightarrow j = 8 \Rightarrow \text{H}
\end{aligned}
$$

The ciphertext is 'JKFPYH', which corresponds to the text obtained with the aid of Table 2.2.   △

An ingenious mechanism in the Hagelin machine determines the value of t and thus selects an alphabet of the Beaufort square for the corresponding encryption. The Hagelin machine contains a drum which is constructed from 27 rods. Two movable teeth are mounted on each rod. The teeth can occupy eight possible positions, two of which inactivate the teeth. The remaining six positions are located opposite the six code wheels. The code wheels are equipped with respectively 26, 25, 23, 21, 19 and 17 pins which are set in either an active or a passive position.

The encryption of a letter is performed during a single revolution of the drum. As the teeth of the drum pass the active pins of the code wheels, the number of passing teeth is registered at the contact points. The resulting value is, in fact, always equal to t and determines which row of the Beaufort square will be used. Before the following letter is encrypted, the code wheels are rotated over a single position, thus moving the active pins, so subsequently different teeth of the drum are counted. The following example demonstrates this process.

Table 2.2. Square of Beaufort.

	A	B	C	D	E	F	G	H	I	J	K	L	M	N	O	P	Q	R	S	T	U	V	W	X	Y	Z
0,26	Z	Y	X	W	V	U	T	S	R	Q	P	O	N	M	L	K	J	I	H	G	F	E	D	C	B	A
1,27	A	Z	Y	X	W	V	U	T	S	R	Q	P	O	N	M	L	K	J	I	H	G	F	E	D	C	B
2	B	A	Z	Y	X	W	V	U	T	S	R	Q	P	O	N	M	L	K	J	I	H	G	F	E	D	C
3	C	B	A	Z	Y	X	W	V	U	T	S	R	Q	P	O	N	M	L	K	J	I	H	G	F	E	D
4	D	C	B	A	Z	Y	X	W	V	U	T	S	R	Q	P	O	N	M	L	K	J	I	H	G	F	E
5	E	D	C	B	A	Z	Y	X	W	V	U	T	S	R	Q	P	O	N	M	L	K	J	I	H	G	F
6	F	E	D	C	B	A	Z	Y	X	W	V	U	T	S	R	Q	P	O	N	M	L	K	J	I	H	G
7	G	F	E	D	C	B	A	Z	Y	X	W	V	U	T	S	R	Q	P	O	N	M	L	K	J	I	H
8	H	G	F	E	D	C	B	A	Z	Y	X	W	V	U	T	S	R	Q	P	O	N	M	L	K	J	I
9	I	H	G	F	E	D	C	B	A	Z	Y	X	W	V	U	T	S	R	Q	P	O	N	M	L	K	J
10	J	I	H	G	F	E	D	C	B	A	Z	Y	X	W	V	U	T	S	R	Q	P	O	N	M	L	K
11	K	J	I	H	G	F	E	D	C	B	A	Z	Y	X	W	V	U	T	S	R	Q	P	O	N	M	L
12	L	K	J	I	H	G	F	E	D	C	B	A	Z	Y	X	W	V	U	T	S	R	Q	P	O	N	M
13	M	L	K	J	I	H	G	F	E	D	C	B	A	Z	Y	X	W	V	U	T	S	R	Q	P	O	N
14	N	M	L	K	J	I	H	G	F	E	D	C	B	A	Z	Y	X	W	V	U	T	S	R	Q	P	O
15	O	N	M	L	K	J	I	H	G	F	E	D	C	B	A	Z	Y	X	W	V	U	T	S	R	Q	P
16	P	O	N	M	L	K	J	I	H	G	F	E	D	C	B	A	Z	Y	X	W	V	U	T	S	R	Q
17	Q	P	O	N	M	L	K	J	I	H	G	F	E	D	C	B	A	Z	Y	X	W	V	U	T	S	R
18	R	Q	P	O	N	M	L	K	J	I	H	G	F	E	D	C	B	A	Z	Y	X	W	V	U	T	S
19	S	R	Q	P	O	N	M	L	K	J	I	H	G	F	E	D	C	B	A	Z	Y	X	W	V	U	T
20	T	S	R	Q	P	O	N	M	L	K	J	I	H	G	F	E	D	C	B	A	Z	Y	X	W	V	U
21	U	T	S	R	Q	P	O	N	M	L	K	J	I	H	G	F	E	D	C	B	A	Z	Y	X	W	V
22	V	U	T	S	R	Q	P	O	N	M	L	K	J	I	H	G	F	E	D	C	B	A	Z	Y	X	W
23	W	V	U	T	S	R	Q	P	O	N	M	L	K	J	I	H	G	F	E	D	C	B	A	Z	Y	X
24	X	W	V	U	T	S	R	Q	P	O	N	M	L	K	J	I	H	G	F	E	D	C	B	A	Z	Y
25	Y	X	W	V	U	T	S	R	Q	P	O	N	M	L	K	J	I	H	G	F	E	D	C	B	A	Z

Figure 2.4. Code wheels and drum of the Hagelin machine.

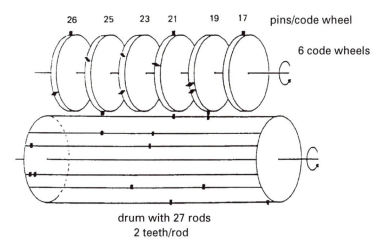

Example

Table 2.3(a) shows the position of the teeth in relation to the six code wheels for each of the 27 rods: a 1 indicates the presence of a tooth at that location, 0 means no tooth. The positions of the active (1) and passive (0) pins on each of the code wheels are given in Table 2.3(b).

Assuming the contact points are set to position 1 on the code wheels, we can see from Table 2.3(b) that only code wheels 3 and 5 have active pins at this position. As the drum and rods are rotated, the number of teeth passing code wheels 3 and 5 is counted. In this example this is equal to $1 + 9 = 10$ (consider the number of ones in the third and fifth rows of Table 2.3(a)). This number determines which row of the Beaufort table ($t = 10$) is used for encryption. For the next letter, the code wheels are rotated one position to position 2, resulting in active pins at positions 1, 4 and 5, as can be seen in Table 2.3(b). Returning to Table 2.3(a) we now find $t = 10 + 3 + 9 = 22$, and so forth. In this manner, all the values of t can be found. However, there is one restriction: two teeth on the same rod will be registered as one single tooth, to ensure that t cannot assume a value larger than 26.

The word CRYPTO is encrypted as follows:

Table 2.3. Hagelin-machine: (a) position of teeth on drum; (b) active pins on code wheels.

(a) rod no.

	1	2	3	4	5	6	7	8	9	10	11	12	13	14	15	16	17	18	19	20	21	22	23	24	25	26	27
1	0	1	1	1	1	0	0	0	0	0	0	0	0	0	0	0	0	0	0	0	0	0	0	0	0	0	0
2	0	0	1	0	0	1	1	1	1	1	1	1	1	1	0	0	0	0	0	0	0	0	0	0	0	0	0
3	0	0	0	0	0	0	0	0	0	0	0	0	0	0	1	0	0	0	0	0	0	0	0	0	0	0	0
4	0	0	0	1	0	0	0	0	0	0	0	0	0	0	1	1	0	0	0	0	0	0	0	0	0	0	0
5	0	0	0	0	0	0	0	0	0	0	0	0	0	0	0	0	1	1	1	1	1	1	1	1	1	0	0
6	0	0	0	0	1	0	0	0	0	1	1	1	1	1	0	1	0	0	0	0	1	1	1	1	1	1	1

code wheel no.

(b) positions on code wheel

	1	2	3	4	5	6	7	8	9	10	11	12	13	14	15	16	17	18	19	20	21	22	23	24	25	26
1	0	0	1	1	1	1	0	1	0	1	1	0	0	0	1	1	0	0	1	0	1	0	0	1	1	1]
2	0	1	1	0	1	0	0	0	1	0	0	1	1	0	0	1	0	1	1	0	1	0	0	0	1]	
3	1	0	0	1	1	1	1	1	1	0	1	0	0	0	1	0	1	0	1	1	0	1	0]			
4	0	1	1	0	1	1	0	0	0	1	0	0	1	0	1	1	0	1	0	1	1]					
5	1	1	1	0	0	1	0	1	0	1	0	0	1	0	1	0	0	0	1]							
6	0	0	1	0	0	1	1	1	0	1	0	1	1	0	1	0	0]									

C: $i = 3$,   t = 10 $\Rightarrow j = 8$ $\Rightarrow$ H

R: $i = 18$,   t = 22 $\Rightarrow j = 5$ $\Rightarrow$ E

Y: $i = 25$,   t = 26 $\Rightarrow j = 2$ $\Rightarrow$ B

P: $i = 16$,   t = 5 $\Rightarrow j = 16$ $\Rightarrow$ P

T: $i = 20$,   t = 15 $\Rightarrow j = 22$ $\Rightarrow$ V

O: $i = 15$,   t = 22 $\Rightarrow j = 8$ $\Rightarrow$ H. $\triangle$

The number of pins on the code wheels has been chosen so that they have no common factors. Therefore it will take no less than $26 \times 25 \times 23 \times 21 \times 19 \times 17 = 101.405.850$ revolutions before all the code wheels have returned to their initial positions. Thus, 101 405 850 letters can be encrypted before the machine repeats the same encryption pattern. This is a maximum value since certain configurations of the active and inactive pins can result in an earlier repetition of the encryption pattern.

The key to the encryption is, in fact, given by the positions of the teeth on the rods and the positions of the pins on the code wheels, so with this knowledge, we can calculate the total number of possibilities for the keys. In all, there are $26 + 25 + 23 + 21 + 19 + 17 = 131$ pins, which can assume one of two settings (active and inactive). Hence, there are 2^{131} possible configurations for the code wheels.

Let us now consider a rod. Each rod has two teeth and zero, one or two teeth are located opposite the six code wheels. There is only one way of locating zero teeth opposite the code wheels and that is by setting both teeth to inactive. There are six ways of placing one tooth opposite the code wheels (either at code wheel 1 or code wheel 2, etc.) and finally, two teeth can be positioned in $\binom{6}{2} = 15$ ways. Therefore, there are $1 + 6 + 15 = 22$ possible ways of fixing two teeth to a rod. Thus, the computation of the total number of configurations of the drum with its 27 rods is equivalent to calculating in how many ways 27 objects can be selected from a set of 22. Obviously, after each selection, the object is returned to the set, as some of the 27 rods will be identical. The solution to this combinatorial problem can be found by drawing the same number of columns as objects and placing a cross × in the corresponding column. For instance, if we wish to select five objects from a set of 4, this may look something like this:

Object 1	Object 2	Object 3	Object 4
× /	/	× × /	× ×

where the delimiter / is used to separate the columns. A shorter notation is ×//×/×. Two slashes directly after one another is interpreted as an object not being selected. Each possibility is entirely defined by a given arrangement of the five crosses and three slashes. The five crosses can be placed in 8 (= 5 + 3) positions in $\binom{8}{5}$ ways. Or, if n objects are selected from a set of m objects, this can be done in $\binom{m+n-1}{n}$ ways. Thus, we find for the total number of possible configurations of the drum:

$$\binom{48}{27} = \frac{48!}{21!\,27!} = 2.23 \times 10^{13}.$$

The total number of keys is therefore:

$$2^{131} \times 2.23 \times 10^{13} = 6.07 \times 10^{52} \text{ keys.}$$

No matter how large this figure may be, the Hagelin machine will eventually fail to withstand cryptanalytic attacks. The six code wheels will assume a certain position, in which the number of active pins is counted. The configuration of the drum, however, will remain unchanged, so that in fact, the letter of the ciphertext is determined by the position of the code wheels. There are $2^6 = 64$ possible positions of the code wheels with respect to the drum. Since 64 cannot be divided by 26, the values for t will exhibit a non-uniform distribution, which is a potential weakness of this method.

A ciphertext-only-attack can provide information on the positions of the active and passive pins of the code wheels. Considering code wheel 6, which has 17 pins, we can write the ciphertext as a matrix, with the first 17 letters in the first row, the second 17 letters in the second, etc. Since code wheel 6 returns to the same position after 17 letters, all the letters in column i will be enciphered by the same active or passive pin. Assuming column i was encrypted by an active pin, then a second column j will show the same distribution if this was also encrypted by an active pin. The same holds for the case in which columns i and j were both encrypted by passive pins. Here we have assumed that the influence of the other code wheels on columns i and j is entirely random. Therefore, by looking at the distributions in the columns, a reasonable idea of which pins are active or passive can be obtained. By arranging the ciphertext in a 19-column matrix, we can estimate which columns are related to the active and passive pins of code wheel 5, and so on for the remaining code wheels.

In practice, 1000–2000 letters of the ciphertext prove sufficient to be able to find the relative positions of the pins of the code wheels. For a known-plaintext-attack, only approximately 50–100 letters are needed.

2.5 Statistics and cryptanalysis

Cryptanalysts has several statistical tools at their disposal for finding the key or plaintext from a ciphertext, or, for instance, for determining whether two columns have the same frequency distribution, as in the case of the Hagelin machine. In this section we will examine several statistical tests which play an important role in cryptanalysis and demonstrate the process of decipherment with a simple example.

Coincidence index (CI)

If we consider a totally random text, constructed from an alphabet of 26 letters, then each letter will have the same probability of occurrence, equal to 1/26. Suppose we have a second random text, which we place beneath the first. We may then wonder how great the chance is of finding two identical letters one above the other. Since each letter exhibits a random character, the probability of finding for example two a's together is equal to $(1/26)^2$. Obviously, this also applies to two b's etc., which results in the total probability of finding two of the same letters together of:

$$(1/26)^2 + (1/26)^2 \ldots + (1/26)^2 = 26 \times (1/26)^2 = 1/26 = 0.0385.$$

However, for an English text, as opposed to a random text, we find that the probabilities of occurrence of the letters are not the same. In English, approximately: $p(a) = 0.082, p(b) = 0.015 \ldots$ etc. Now, the calculation of the probability of finding two identical letters together yields:

$$(0.082)^2 + (0.015)^2 + \ldots = 0.0661.$$

This value is larger than in the case of a random text. This is referred to as the coincidence index and is generally defined according to the following expression:

$$\text{CI} = \sum_{i=1}^{n} p_i^2, \tag{2.1}$$

in which n is the size of the alphabet and p_i the probability of occurrence of the ith symbol of the alphabet.

In the preceding section we found that for a random and an English text the CI was equal to 0.0385 and 0.0661, respectively. Every language is characterised by a specific value of CI, as is shown in Table 2.4.

In cryptanalysis, the calculation of CI can prove worthwhile in several ways. In the case of a monoalphabetical substitution the letters of the plaintext are replaced by other letters. This does not influence the statistical parameters of the text and therefore neither the value of CI of the plaintext nor the ciphertext. We can use this information to test whether a text was enciphered by means of a monoalphabetical or a polyalphabetical substitution. If the CI value of a text corresponds roughly to that of the language in which the text is written, so that:

$$CI(plaintext) = CI(ciphertext),$$

then it is likely that a monoalphabetical substitution has been used. However, if the value of CI of the ciphertext turns out to be considerably lower, a polyalphabetical substitution could have been used. Since a polyalphabetical substitution will tend to conceal the statistical parameters of a text the value of CI will approach that of a random text.

In addition, a calculation of the CI can also provide insight into the probability of two different ciphertexts, say C_1 and C_2, being encrypted by the same method. If this is the case, then $CI(C_1) \approx CI(C_2)$. At the end of this section, we will present an example which relies on this principle.

Since in practice ciphertexts have only a finite length, the value we find for the CI from the ciphertext will always differ from the theoretical value. For this reason, we often use an estimation, CI′, instead of CI. A suitable choice of CI′ is given by:

$$CI' = \sum_{i=1}^{n} x_i(x_i - 1)/L(L - 1) \tag{2.2}$$

in which L represents the length of the ciphertext and x_i the number of occurrences of symbol i in the ciphertext. The probability that in a ciphertext of length L, symbol i will occur x_i times can be described by a binomial distribution:

Table 2.4. Values of the CI for various languages.

English	0.0661
French	0.0778
German	0.0762
Italian	0.0738
Japanese	0.0819
Russian	0.0529
random text	0.0385

$$p(x_i) = \binom{L}{x_i} p_i^{x_i}(1 - p_i)^{L-x_i}. \tag{2.3}$$

Assuming L is sufficiently large, we can derive the following expressions for the expectation and the variance of x_i:

$$E(x_i) = Lp_i, \tag{2.4}$$

$$\mathrm{var}(x_i) = Lp_i(1 - p_i). \tag{2.5}$$

These two expressions can be used to prove that CI$'$ is a pure estimator of CI.

Theorem 2.1

Let the CI be defined as:

$$CI = \sum_{i=1}^{n} p_i^2,$$

and let

$$CI' = \sum_{i=1}^{n} x_i(x_i - 1)/L(L - 1),$$

then CI$'$ is a pure estimator of CI. This implies that

$$E(CI') = CI. \tag{2.6}$$

Proof

We can write:

$$E(CI') = E\left[\sum_i x_i(x_i - 1)/L(L - 1)\right]$$

$$= \sum_i E[x_i(x_i - 1)/L(L - 1)].$$

Since $\mathrm{var}(x_i) = E(x_i^2) - [E(x_i)]^2$ we find for $E(CI')$:

$$E(CI') = \sum_i E\left[(x_i^2 - x_i)/L(L - 1)\right]$$

$$= \sum_i [E(x_i^2) - E(x_i)]/L(L - 1)$$

$$= \sum_i \{\mathrm{var}(x_i) + [E(x_i)]^2 - E(x_i)\}/L(L - 1).$$

Substitution of the previously found formulae for $E(x_i)$ and $\text{var}(x_i)$, eqs. (2.4) and (2.5), respectively, in this expression, leads to the following result:

$$E(CI') = \sum_i [Lp_i(1 - p_i) + L^2p_i^2 - Lp_i]/L(L - 1)$$

$$= \sum_i [Lp_i - Lp_i^2 + L^2p_i^2 - Lp_i]/L(L - 1)$$

$$= \sum_i p_i^2 = CI,$$

which is exactly the result we were looking for. $\square$

Kasiski test

It may happen that at different places in the ciphertext identical sequences of letters appear. These repeated patterns are interesting as they can provide information on periodicity within the text. Consider the following example, in which the plaintext is transformed to a ciphertext with a given key:

plaintext: REQUESTS ADDITIONAL TEST ...
key: TELEXTEL EXTELEXTEL EXTE ...
ciphertext: CAVKTBLT EUQWSWJGEA LTBL ...

The plaintext contains the letter sequence EST twice. Since for both cases, the same section of the key is used for encryption, the resulting letter sequence TBL in the ciphertext is also the same for both cases. This is caused by the fact that the sequences EST are positioned exactly a multiple number of the key length, or period, apart. Clearly, the distance between identical letter sequences can tell us something about the period of an encrypted text. We can find this value by determining the most frequently occurring common factor.

Example
A given ciphertext contains letter sequences repeated at the following distances:

		distance
PQA	150	$= 2 \times 5^2 \times 3$
RET	42	$= 2 \times 7 \times 3$
FRT	10	$= 2 \times 5$
ROPY	81	$= 3^4$

DER $57 = 19 \times 3$
RUN $117 = 13 \times 3^2$

Since a factor 3 appears as the most common factor, we can state that the period of the ciphertext is most probably 3.   △

Chi test

The chi-test offers a straightforward means of comparing two frequency distributions. The following sum is calculated, in which p_i represents the uncertainty of the occurrence of symbol i with the first distribution and q_i the uncertainty for the second distribution.

$$\chi = \sum_{i=1}^{n} p_i q_i. \tag{2.7}$$

It is evident that when the two frequency distributions are similar, the value of χ will be higher than when the two distributions are dissimilar.

Assume we have two ciphertexts C_1 and C_2, which are both the result of a Caesar substitution. The alphabet of the first is shifted by t_1 letters; that of the second by t_2. If $t_1 = t_2$, i.e. C_1 and C_2 are encrypted with the same Caesar substitution, then χ will be large, since the statistics of C_1 will exhibit large similarities with those of C_2. Correspondingly, when $t_1 \neq t_2$, χ will be small.

Besides being used to determine whether the same or different substitutions have been employed, χ can also be used to reduce a poly-alphabetical substitution to a monoalphabetical substitution. This is illustrated by the following example.

Example

A given plaintext is transposed to a ciphertext with the Vigenère table and the key-word RADIO:

plaintext: EXECUTE THESE COMMANDS
key: RADIORA DIORA DIORADIO
ciphertext: VXHKIKE WPSJE FWADAQLG

In order to convert the ciphertext back to the plaintext, it is written in a matrix whose number of columns corresponds to the length of the key word:

R	A	D	I	O
V	X	H	K	I
K	E	W	P	S
J	E	F	W	A
D	A	Q	L	G

The original plaintext can be retrieved by deciphering the first column with row *R* of the Vigenère table, the second column with row *A*, the third with row *D*, etc.

The decipherment can also be performed in a different way. Consider the letters of the key-word and their relative distances to the first letter, in this case the letter R. This yields the following values:

R	A	D	I	O
0	9	12	17	23

By replacing the letters of column 2 of the ciphertext with the letters which are located 9 positions earlier in the alphabet, the letters of column 3 with letters which are located 12 places earlier, etc., we obtain the following table:

V	O	V	T	L
K	V	K	Y	V
J	V	T	F	D
D	R	E	U	J

Now only row *R* of the Vigenère table is used for deciphering the entire ciphertext, instead of five different rows of the table.

We have now, in fact, reduced the problem of decrypting a ciphertext based on a polyalphabetical substitution to deciphering a text based on a monoalphabetical substitution. $\triangle$

This procedure for converting a polyalphabetical substitution to a mono-alphabetical substitution is only effective when the distances between the letters of the key word are known. Usually, though, a cryptanalyst does not have this information. The chi test, however, offers a means of finding some indication of these distances. In the above example we saw that the letters of the columns of the ciphertext were shifted such that they could all be deciphered with one and the same substitution cipher. Since the columns are encrypted with the same monoalphabetical substitution, the value of χ calculated for two columns will be high. Now all that remains for the cryptanalyst is to shift the letters of a column repeatedly, until the highest value for χ calculated for a given column and the first one is found. It is then most likely that the columns can be deciphered with one and the same key row.

This section has given several examples of the statistical tools which are available to the cryptanalyst. Finally we will now give an example of a practical analysis of a ciphertext.

An example of cryptanalysis

Two ciphertexts are intercepted, shortly after one another:

Cipher text 1:
```
k o o m m a c o m o q e g l x x m q c c k u e y f c u r
y l y l i g z s x c z v b c k m y o p n p o g d g i a z
t x d d i a k n v o m x h i e m r d e z v x b m z r n l
z a y q i q x g k k k p n e v h o v v b k k t c s s e p
k g d h x y v j m r d k b c j u e f m a k n t d r x b i
e m r d p r r j b x f q n e m x d r l b c j h p z t v v
i x y e t n i i a w d r g n o m r z r r e i k i o x r u
s x c r e t v
```

Cipher text 2:
```
z a o z y g y u k n d w p i o u o r i y r h h b z x r c
e a y v x u v r x k c m a x s t x s e p b r x c s l r u
k v b x t g z u g g d w h x m x c s x b i k t n s l r j
z h b x m s p u n g z r g k u d x n a u f c m r z x j r
y w y m i
```

As these two ciphertexts were received within a short time of each other, it is very possible that they were both enciphered according to the same method. This hypothesis can be verified by calculating CI′ for both texts:

$$CI'(C_1) = 0.0421,$$

$$CI'(C_2) = 0.0445.$$

These values correspond well enough to assume safely that both texts were indeed enciphered according to the same method.

The values of CI′ lie somewhere between that of a random text and, for instance, an English text. This would indicate a polyalphabetical substitution. In order to determine whether this is actually the case, we can use the Kasiski test, for which we must find repeated letter sequences in the ciphertext and their distances. The results are given below:

Kasiski test ciphertext 1

ak	70
bc	35, 84
cj	35
ck	22
dr	21, 22
em	13, 70
et	29
gd	63
ia	7, 115
ie	70
kk	11, 12
kn	70
ma	126
mr	21, 42, 49
mx	88
ne	56
om	5, 58, 117
pn	49
rd	21, 49
re	12
rr	41
sx	161
tv	36
ue	106
vb	63
vv	65
xb	60
xc	161
xd	98
xy	53
yl	2
zr	105
zt	109
zv	37
akn	70
bcj	35
emr	70
iem	70
mrd	21, 49
sxc	161
emrd	70
iemr	70
iemrd	70

Kasiski test ciphertext 2

bx	28
cm	67
cs	21
dw	56
gz	32
hb	63
lr	28
rx	14
sl	28
uk	48
xc	21
xm	18
xs	3
zx	84
slr	28
xcs	21

When these distances are resolved as products of prime numbers, the number 7 appears relatively frequently. It is therefore probable that the period is equal to 7 and that 7 alphabets were used for encryption. We can now write the ciphertexts in matrix form.

```
k o o m m a c
o m o q e g l
x x m q c c k
u e y f c u r
y l y l i g z
s x c z v b c
k m y o p n p
o g d g i a z
t x d d i a k
n v o m x h i
e m r d e z v
x b m z r n l
z a y q i q x
g k k k p n e
v h o v v b k
k t c s s e p
k g d h x y v
j m r d k b c
j u e f m a k
n t d r x b i
e m r d p r r
j b x f q n e
m x d r l b c
j h p z t v v
i x y e t n i
i a w d r g n
o m r z r r e
i k i o x r u
s x c r e t v
z a o z y g y
u k n d w p i
o u o r i y r
h h b z x r c
e a y v x u v
r x k c m a x
s t x s e p b
r x c s l r u
k v b x t g z
u g g d w h x
m x c s x b i
k t n s l r j
z h b x m s p
u n g z r g k
u d x n a u f
c m r z x j r
y w y m i
```

Each column should correspond to a certain monoalphabetical substitution. Calculation of the values CI' for the various columns results in:

$$CI'(\text{column } 1) = 0.0522,$$
$$CI'(\text{column } 2) = 0.0801,$$
$$CI'(\text{column } 3) = 0.0734,$$
$$CI'(\text{column } 4) = 0.0744,$$
$$CI'(\text{column } 5) = 0.0705,$$
$$CI'(\text{column } 6) = 0.0717,$$
$$CI'(\text{column } 7) = 0.0606.$$

As far as the values of CI′ are concerned, it seems likey that each column has been enciphered with a monoalphabetical substitution. We can now attempt to convert the ciphertext based on a polyalphabetical substitution to a different ciphertext which can be deciphered with a single monoalphabetical substitution. First, we repeatedly shift the letters of each column, starting at position 1 and ending with position 25, and each time we calculate the corresponding value of χ, with respect to another column. At a certain point, one of these values of χ is found to be considerably higher than the others. These have been underlined below:

columns 1 and 2:   0.0388 0.0487 0.0317 0.0326 0.0274 0.0340 0.0421 0.0402 0.0321
0.0350 0.0425 0.0411 <u>0.0662</u> 0.0350 0.0317 0.0359 0.0491 0.0331
0.0236 0.0378 0.0345 0.0288 0.0567 0.0525 0.0302

columns 1 and 3:   0.0378 0.0274 0.0331 0.0331 0.0250 0.0581 0.0491 0.0458 0.0284
0.0383 0.0529 0.0491 0.0307 0.0250 0.0312 0.0444 0.0392 0.0359
0.0392 0.0354 0.0421 <u>0.0657</u> 0.0416 0.0269 0.0232

columns 1 and 4:   0.0317 0.0369 0.0364 0.0312 0.0454 0.0383 <u>0.0558</u> 0.0302 0.0388
0.0345 0.0520 0.0250 0.0359 0.0336 0.0477 0.0260 0.0435 0.0520
0.0406 0.0369 0.0468 0.0468 0.0326 0.0307 0.0326

columns 1 and 5:   0.0430 <u>0.0586</u> 0.0279 0.0274 0.0331 0.0473 0.0298 0.0359 0.0336
0.0354 0.0302 0.0491 0.0548 0.0265 0.0359 0.0406 0.0506 0.0312
0.0345 0.0336 0.0440 0.0354 0.0506 0.0411 0.0321

columns 1 and 6:   0.0382 0.0271 0.0502 0.0449 0.0295 0.0319 0.0440 0.0522 0.0372
0.0372 0.0343 0.0396 0.0391 0.0391 0.0280 0.0324 0.0454 0.0430
<u>0.0614</u> 0.0362 0.0324 0.0343 0.0498 0.0338 0.0275

columns 1 and 7:   0.0285 0.0444 0.0362 0.0382 0.0357 0.0353 0.0343 0.0415 0.0377
0.0483 0.0333 0.0396 0.0425 0.0300 <u>0.0565</u> 0.0348 0.0329 0.0348
0.0454 0.0304 0.0377 0.0324 0.0449 0.0295 0.0444

columns 2 and 3:   0.0156 0.0369 0.0288 0.0350 0.0340 0.0506 0.0222 0.0350 <u>0.0827</u>
0.0440 0.0307 0.0302 0.0340 0.0345 0.0340 0.0364 0.0307 0.0293
0.0373 0.0402 0.0600 0.0416 0.0378 0.0477 0.0558

columns 2 and 4:   0.0411 0.0411 0.0321 0.0317 0.0473 0.0317 0.0548 0.0421 0.0562
0.0279 0.0373 0.0227 0.0435 0.0274 0.0402 0.0397 0.0312 0.0336
0.0255 <u>0.0775</u> 0.0425 0.0369 0.0284 0.0629 0.0142

columns 2 and 5:   0.0284 0.0435 0.0369 0.0586 0.0227 0.0364 0.0274 0.0463 0.0340
0.0406 0.0473 0.0336 0.0288 0.0298 <u>0.0822</u> 0.0293 0.0321 0.0288
0.0454 0.0161 0.0458 0.0411 0.0345 0.0336 0.0345

columns 2 and 6:   0.0290 0.0295 0.0377 0.0300 0.0343 <u>0.0874</u> 0.0304 0.0295 0.0304
0.0473 0.0329 0.0401 0.0440 0.0314 0.0232 0.0338 0.0444 0.0304
0.0464 0.0464 0.0478 0.0420 0.0478 0.0179 0.0493

columns 2 and 7:   0.0251 <u>0.0700</u> 0.0348 0.0411 0.0319 0.0420 0.0150 0.0454 0.0353
0.0425 0.0396 0.0401 0.0502 0.0174 0.0589 0.0309 0.0440 0.0391
0.0377 0.0169 0.0527 0.0377 0.0329 0.0473 0.0406

columns 3 and 4:   0.0312 0.0269 0.0444 0.0321 0.0321 0.0369 0.0463 0.0236 0.0336
0.0411 <u>0.0676</u> 0.0440 0.0298 0.0397 0.0444 0.0246 0.0269 0.0440
0.0279 0.0326 0.0383 0.0421 0.0331 0.0468 0.0671

columns 3 and 5:   0.0454 0.0444 0.0232 0.0326 0.0529 <u>0.0662</u> 0.0345 0.0246 0.0449
0.0336 0.0312 0.0430 0.0454 0.0203 0.0364 0.0473 0.0548 0.0260
0.0350 0.0572 0.0406 0.0265 0.0241 0.0378 0.0317

columns 3 and 6:   0.0435 0.0425 0.0314 0.0420 0.0237 0.0295 0.0449 0.0444 0.0309
0.0367 0.0594 0.0386 0.0382 0.0367 0.0343 0.0430 0.0377 0.0300
0.0271 0.0213 0.0338 0.0464 <u>0.0787</u> 0.0348 0.0261

columns 3 and 7:   0.0338 0.0396 0.0411 0.0507 0.0251 0.0459 0.0502 0.0333 0.0411
0.0227 0.0304 0.0401 0.0444 0.0348 0.0478 0.0338 0.0319 0.0367
<u>0.0556</u> 0.0449 0.0304 0.0430 0.0338 0.0266 0.0401

columns 4 and 5:   0.0402 0.0558 0.0298 0.0312 0.0260 0.0704 0.0288 0.0548 0.0312
0.0534 0.0147 0.0317 0.0435 0.0487 0.0321 0.0232 0.0577 0.0222
0.0378 0.0350 <u>0.0789</u> 0.0189 0.0359 0.0236 0.0435

columns 4 and 6:   0.0459 0.0324 0.0406 0.0406 0.0502 0.0295 0.0179 0.0459 0.0251
0.0396 0.0401 <u>0.0734</u> 0.0232 0.0319 0.0271 0.0541 0.0396 0.0386
0.0415 0.0246 0.0242 0.0304 0.0536 0.0459 0.0565

columns 4 and 7:   0.0488 0.0411 0.0295 0.0522 0.0261 0.0401 0.0256 <u>0.0720</u> 0.0285
0.0469 0.0188 0.0372 0.0295 0.0435 0.0493 0.0343 0.0430 0.0256
0.0478 0.0285 0.0638 0.0271 0.0556 0.0242 0.0251

columns 5 and 6:   0.0237 0.0546 0.0343 0.0449 0.0285 0.0556 0.0295 0.0382 0.0290
0.0536 0.0444 0.0222 0.0367 0.0300 0.0338 0.0353 <u>0.0768</u> 0.0285
0.0304 0.0251 0.0546 0.0454 0.0435 0.0425 0.0348

columns 5 and 7:   0.0256 0.0580 0.0304 0.0333 0.0242 0.0478 0.0430 0.0324 0.0478
0.0290 0.0367 0.0217 <u>0. 0739</u> 0.0300 0.0502 0.0227 0.0430
0.0300 0.0372 0.0454 0.0430 0.0396 0.0237 0.0507 0.0213

columns 6 and 7:   0.0212 0.0420 0.0405 0.0400 0.0489 0.0326 0.0479 0.0212 0.0533
0.0311 0.0479 0.0365 0.0267 0.0242 0.0440 0.0484 0.0370 0.0553
0.0351 0.0272 0.0212 <u>0.0652</u> 0.0341 0.0474 0.0385

Based on these results, the most probable distances across which the columns have been shifted with respect to column 1 are:

> column 2: 13
> column 3: 22
> column 4: 7
> column 5: 2
> column 6: 19
> column 7: 15

These values are then used to convert ciphertexts C_1 and C_2 to the following text:

```
k b k t o t r o z k x g z a x k i x e
v z u r u m e n g y y u s k z o s k y
g x u r k z u v r g e o t z n k t o t
k z k k t z n i k t z a x e z n k g s
k x o i g t g a z n u x k j m g x g r
r g t v u k c x u z k g y z u x e k t
z o z r k j z n k m u r j h a m o t z
n g z y z u x e z n k r k g j o t m s
g t m k z y n u r j u l g v o k i k u
l v g x i n s k t z c o z n g t k t i
x e v z k j s k y y g m k z n k g a z
n u x j k y i x o h k y k r g h u x g
z k r e n u c z n k r k g j o t m s g
t z g i q r k y z n k j k i x e v z o
u t c k y a m m k y z z u x k g j z n
k y z u x e o l e u a c g t z z u q t
u c n u c z n g z c g y j u t k
```

It should now be possible to decipher this text with a single mono-alphabetical substitution. It is left to the reader to investigate this and to find the plaintext!   △

3

The information theoretical approach

3.1 The general scheme

Figure 3.1 shows the general scheme of a cipher system. To the left, a source generates a message, also called the plaintext, which is denoted by the letter M. The plaintext is converted to a ciphertext, which is indicated by the letter C, by means of an encipherment algorithm. Encryption can be regarded as a transformation T, which converts M to C. The ciphertext C may be the result of a transposition or substitution cipher, although naturally other methods are also possible. The specific transformation depends on the key K, which is chosen from the set of all conceivable keys. Therefore, we can write:

$$C = T_K(M).$$

At the receiver, the ciphertext is decrypted by performing the inverse transformation T_K^{-1}, so $M = T_K^{-1}(C)$. In most cases, the transformation T

Figure 3.1. General scheme of a cipher system.

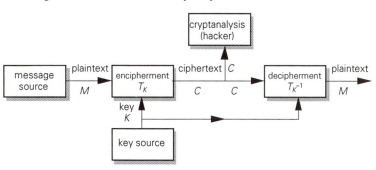

itself is known and only the key is secret. Therefore, the task of the crypt-analyst in the figure will focus on finding the correct key. In this chapter, we will assume that the key is different for every new message and consequently that the plaintext, the ciphertext and the key can be considered as stochastic quantities.

3.2 The information measure and absolute security

A successful application of cryptographic techniques requires insight into the security of the system employed. In this section we will use the terminology of the field of information theory to describe a secure cipher system.[1]

We can speak of a measure of information of the plaintext if we consider the plaintext as a subset of elements, selected from the set of all possible messages, where each element has its own given uncertainty. An expression for the measure of information of the plaintext is:

$$H(M) = -\sum_{i=1}^{n} p(M_i) \log p(M_i),\tag{3.1}$$

in which $p(M_i)$, $i = 1,...,n$ represent the uncertainty of occurrence of the possible plaintexts, under the assumption that the texts are mutually independent.

In the same manner, we can introduce a measure of the information of the ciphertext, denoted by $H(C)$, and a measure of information for the keys, $H(K)$, which are defined as:

$$H(C) = -\sum_{j=1}^{m} p(C_j) \log p(C_j),\tag{3.2}$$

and

$$H(K) = -\sum_{h=1}^{\ell} p(K_h) \log p(K_h),\tag{3.3}$$

in which C_j , $j = 1,...,m$ represent the possible ciphertexts and K_h, $h = 1,...,\ell$ the possible keys. Note that due to the one-to-one relation between the ciphertext and plaintext, it must hold that $m = n$.

---

[1] Those who are not yet acquainted with the fundamentals of information theory are referred to Appendix A.

Correspondingly, we can also speak of the *conditional information measure, H(K/C)*. This is a measure of the information or uncertainty with respect to the key when the ciphertext C is available. It is also referred to as *key equivocation* and is defined as follows:

$$H(K/C) = -\sum_{h=1}^{\ell} \sum_{j=1}^{m} p(K_h, C_j) \log p(K_h/C_j). \tag{3.4}$$

$H(M/C)$ is the measure of information or uncertainty with respect to the plaintext M, for a given ciphertext C, and is also called the *message equivocation*:

$$H(M/C) = -\sum_{i=1}^{n} \sum_{j=1}^{m} p(M_i, C_j) \log p(M_i/C_j). \tag{3.5}$$

$H(K/M,C)$ is the measure of information or uncertainty with respect to the key, for a given ciphertext and corresponding plaintext.

$$H(K/M,C) = -\sum_{h=1}^{\ell} \sum_{i=1}^{n} \sum_{j=1}^{m} p(K_h, M_i, C_j) \log p(K_h/M_i, C_j). \tag{3.6}$$

This quantity is often also referred to as the *key appearance equivocation*. In the same manner, $H(M/C,K)$ is a measure of the information or uncertainty with respect to the plaintext when both the ciphertext and the key are known. Since there is a one-to-one relation between the plaintext and the ciphertext with a given key, the following must always hold:

$$H(M/C,K) = 0. \tag{3.7}$$

Naturally, when both the ciphertext and the key are available we can retrieve the plaintext without any difficulty whatsoever: the uncertainty with respect to M is zero. The receiver will therefore suffer no loss of information; after it has been deciphered, the entire original text is known.

The relations between the three expressions for the equivocation are stated below:

Theorem 3.1
Let $H(K/C)$, $H(M/C)$ and $H(K/M,C)$ be defined as in eqs. (3.4), (3.5) and (3.6), respectively, then:

$$H(K/M,C) = H(K/C) - H(M/C). \tag{3.8}$$

Proof

Information theory teaches us that the measure of information in the plaintext, ciphertext and key is:

$$H(M,C,K) = H(M/C,K) + H(C,K) = H(K/M,C) + H(M,C). \quad (3.9)$$

Remembering that also:

$$H(C,K) = H(K/C) + H(C)$$

and

$$H(M,C) = H(M/C) + H(C),$$

we find that:

$$H(M/C,K) + H(K/C) = H(K/M,C) + H(M/C). \quad (3.10)$$

Referring to eq. (3.7), we have already seen that $H(M/C,K) = 0$. This enables us to rewrite eq. (3.10) in the same form as eq. (3.8). □

We can draw several interesting conclusions from Theorem 3.1. Someone using a cipher system will always strive for a high value of $H(K/M,C)$. If a cryptanalyst has succeeded in obtaining both the plaintext and the ciphertext, then the uncertainty with respect to the key must be as large as possible. We can see from eq. (3.8) that in order to obtain a large value for $H(K/M,C)$, $H(M/C)$ must be small. However, a small value for $H(M/C)$ implies that little uncertainty remains with respect to the plaintext when the ciphertext alone is available, which is precisely what we had hoped to avoid. The larger the uncertainty with respect to the key, the smaller the uncertainty with respect to the plaintext and vice versa, the larger the uncertainty with respect to the plaintext, the smaller the uncertainty with respect to the key.

The information theoretical approach leads to additional conclusions. Let $I(M;C)$ represent the mutual information of the plaintext and the ciphertext, which is defined as:

$$I(M;C) = H(M) - H(M/C) = H(C) - H(C/M). \quad (3.11)$$

The objective of someone implementing a cipher system will be to minimise $I(M;C)$. If the ciphertext provides no information about the plaintext, then $H(M/C) = H(M)$. With eq. (3.11), we then find that the mutual information between the plaintext and the ciphertext is equal to zero. We call this an *absolutely secure crypto-system*, i.e.:

$$I(M;C) = 0. \quad (3.12)$$

One-time pad

An example of an absolutely secure cipher system is provided by the *one-time pad*, which was introduced in 1926 by G. S. Vernam. Our description of this system will be based on the English alphabet, with its 26 letters, but without commas, periods, spaces, etc. The message $M = (m_1,...,m_L)$ has a total length of L letters. In order to encipher M, a random sequence of L letters from the same alphabet is generated. Each letter has a probability of 1/26 of being selected and each selection is independent of the others. This series of letters $(k_1,...,k_L)$ is used as the key to encrypt M. The ciphertext $C = (c_1,...,c_L)$ is obtained by calculating

$$c_i = m_i + k_i \text{ (mod 26)} \tag{3.13}$$

for every $i = 1,...,L$. Each letter is represented by a numerical value {0,1,...,25}, which corresponds to its position in the alphabet. The term (mod 26) ensures the left- and right-hand sides of the equation are equal, save for a multiple of 26.

It is clear therefore that there are 26^L possible keys with the same uncertainty and hence

$$H(K) = L \log 26.$$

Since K consists of independently chosen letters from the alphabet and for every m_i and c_i a unique k_i exists, then for every combination of M and C:

$$p(C/M) = 1/26^L.$$

The total number of possibilities for the ciphertext is also 26^L, where each possibility has a probability of occurrence of:

$$p(C) = 1/26^L.$$

Thus, M and C are independent and $I(M;C) = 0$. Clearly, the one-time pad is an absolutely secure crypto-system. The term 'one-time' refers to the fact that in order to guarantee the security of the system, each key may be used only once. Before each message is transmitted a new key must be generated. In practice, the fact that every message must be accompanied by a key of the same length can be a disadvantage, as the key must be transmitted before the message. Despite this disadvantage, the one-time pad is still used for applications which require a high level of security, for instance the hot-line between Washington and Moscow.

The following theorem provides a lower limit for the mutual information with respect to the plaintext and the ciphertext.

Theorem 3.2

$$I(M;C) \geq H(M) - H(K). \tag{3.14}$$

Proof

We will start our proof of this theorem by first considering eq. (3.8). Since $H(K/M,C) \geq 0$, it follows that

$$H(K/C) \geq H(M/C). \tag{3.15}$$

As

$$H(K) \geq H(K/C),$$

eq. (3.15) leads to

$$H(K) \geq H(M/C).$$

When this expression is substituted in eq. (3.11), eq. (3.14) immediately follows.   □

Theorem 3.2 implies that when the uncertainty of a set of keys is small, the mutual information will be large (on average) and therefore so will independence of the plaintext and the ciphertext.

Absolute security of a system, i.e. $I(M;C) = 0$, can only be achieved when

$$H(K) \geq H(M). \tag{3.16}$$

Assuming that each key and each plaintext has the same probability of occurrence, eq. (3.16) can be rewritten as:

$$K \geq M. \tag{3.17}$$

We have already seen that the one-time pad satisfies this condition.

The conditions for an absolutely secure cipher system, as stated in eqs. (3.16) and (3.17), are not very encouraging, since the key must be at least the same length as the plaintext.

However, in 1989 Maurer (Maurer, 1989, Massey, 1990) demonstrated that the pessimistic inequality of eq. (3.16) can be avoided by introducing the concept of *security events*. By definition, when a security event S occurs, the system can be considered an absolutely secure system. When S does not occur, the cipher system may not be entirely secure. Maurer demonstrated that even when there is a small chance of a security failure, an otherwise absolutely secure cipher system is still conceivable, for which $H(K) < H(M)$. We will illustrate this with the following example.

Consider a telephone network with 2^L telephones. Each telephone number consists of L bits. When a call is made to telephone I, it replies by generating a random binary sequence R_I of N bits (with $N \gg L$), which is unique for that telephone number. The secret key K is made up of a secret telephone number of L bits and one of the 2^L numbers of the network. This key is known to the transmitter and receiver, but obviously not to an intruder. The system operates as follows (see also Figure 3.2). When A wishes to transmit a plaintext M of N bits to B, A first dials the secret number K and receives a random sequence R_K. Then A transmits a ciphertext to B, which is the result of the following addition of M and R_K:

$$C = M + R_K \ (\text{mod } 2). \tag{3.18}$$

This is similar to the one-time pad. When B has received C, R_K is also obtained by dialling K, in order to retrieve the plaintext from C.

The security event S of this example is the fact that an intruder who has managed to intercept C and tries to retrieve M, will not dial the secret number K. Without dialling K, the sequence R_K remains completely unpredictable to an intruder. In the case of S, the system is an absolutely secure system. However, if the intruder accidentally calls K, he can decipher C. In this case the system has become unreliable.

If the intruder dials t telephone numbers at random, the probability of S not occurring is given by $p(\text{not-}S) = t/2^K$. Say, for example, $K = 100$ and the intruder tries dialling $t = 2^{50} \approx 10^{15}$ different numbers. Then $p(\text{not-}S) =$

Figure 3.2. Cipher system with security events.

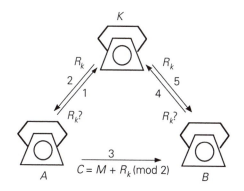

$2^{50}/2^{100} \approx 10^{-15}$. The chance of the intruder finding R_K and the plaintext M is obviously extremely small.

Note that in this system N (which is the length of the plaintext and the random sequence R_K) can be many times larger than the key K: $N \gg K$. In other words, $H(M)$ can be larger than $H(K)$, in which case the expressions (3.16) and (3.17) no longer hold.

The introduction of the concept of security events offers a new way of regarding the security of a system based on information theoretical tools. Absolute security can be guaranteed, even when the length of the key is smaller than that of the plaintext, provided that a security event can be defined for the system and that this event actually occurs.

3.3 The unicity distance

In the previous section we did not take into consideration the importance of the length of the ciphertext, when this is intercepted by a cryptanalyst. Suppose the plaintext was written in ordinary, spoken language. We have already seen that this kind of text will exhibit certain statistical characteristics, which the cryptanalyst can use in attempting to crack the ciphertext.

The chances of a cryptanalyst succeeding in deciphering a text generally increase as the length of the text increases. Denoting a ciphertext of length L by C^L, the key equivocation $H(K/C^L)$ will decrease for increasing values of L. The uncertainty with respect to the key used will diminish as L increases. This is depicted in Figure 3.3. The key equivocation can even reach zero when the key can be determined from the ciphertext. The maximum value of the key equivocation is $H(K)$. This value occurs when the uncertainty of a key is equal for all keys, with a given ciphertext. In this case, maximum uncertainty is reached, with respect to the key used.

A similar line of thought applies to the message equivocation $H(M^L/C^L)$, although there is one exception: when L is small, the number of possible messages is also small and therefore the message equivocation remains low. As soon as L increases, the number of possible messages increases dramatically and, consequently, the message equivocation rises. Once L is sufficiently large, the ciphertext will contain enough information and the number of most probable messages will no longer increase. Soon after this point is reached, the message equivocation curve coincides with that of key equivocation, because now, the ciphertext contains sufficient information to retrieve the key from the found plaintext with the same uncertainty and vice versa. This is also illustrated in Figure 3.3.

Finally, the figure also shows the curve of the key appearance equivocation $H(K/C^L,M^L)$. Clearly the key appearance equivocation will decrease to zero more quickly than the key equivocation, since in this case we assume the cryptanalyst possesses both the plaintext and the ciphertext and will therefore be better equipped to find the key quickly. The key appearance equivocation is a measure of the strength of a cipher which is exposed to a known-plaintext-attack on the key, whereas the key equivocation and message equivocation are measures of the strength of a cipher exposed to a ciphertext-only-attack on the key and the plaintext.

From the above it is clear that the longer received ciphertext is, the greater is the probability that a cryptanalyst will find the key or the plaintext. We can therefore wonder how large L must be in order to be certain that a cryptanalyst will find the key. First, though, we will consider the following theorem.

Theorem 3.3

The key equivocation $H(K/C^L)$ has the following lower limit

$$H(K/C^L) \geq H(K) - D_L, \tag{3.19}$$

in which D_L represents the redundancy of the plaintext.

Proof

Due to the one-to-one relation between the plaintext and the ciphertext, the following expression will always be satisfied:

Figure 3.3. Key, message and key appearance equivocation as a function of L.

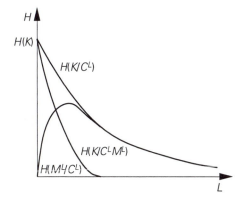

$$H(K,C^L) = H(K,M^L).$$

By assuming the key is independent of the plaintext we can derive for the key equivocation that:

$$H(K/C^L) = H(K,C^L) - H(C^L)$$

$$= H(K,M^L) - H(C^L)$$

$$= H(K) + H(M^L) - H(C^L). \qquad (3.20)$$

Suppose that the number of different symbols of a plaintext or ciphertext is ε. Then the total number of possibilities for plaintexts of length L is ε^L. With the aid of information theory we can prove that always

$$H(C^L) \le L \log(\varepsilon).$$

Eq. (3.20) can therefore be rewritten as

$$H(K/C^L) \ge H(K) + H(M^L) - L \log(\varepsilon). \qquad (3.21)$$

The term $-[H(M^L) - L \log(\varepsilon)]$ in eq. (3.21) is referred to as the redundancy and is represented by D_L.

$$D_L = L \log(\varepsilon) - H(M^L). \qquad (3.22)$$

It is a measure of the discrepancy between the actual source producing messages with a certain probability distribution and a source which produces all messages with equal probability.

Substitution of eq. (3.22) in eq. (3.21) yields eq. (3.19).   □

We can interpret Theorem 3.3 by saying that as the redundancy increases, the average key equivocation and thus the uncertainty with respect to the key used, decreases. Or, in other words: redundancy makes the task of finding the key easier. Conversely, methods for reducing the redundancy can improve the security of a crypto-system.

Another interpretation of the theorem is that as long as it holds that

$$H(K) > D_L,$$

the key equivocation will not become zero and that therefore, generally, the key cannot be determined unequivocally.

The following theorem states under which conditions for L the key equivocation can become zero.

Theorem 3.4

When the plaintext is generated by a source whose successive symbols are mutually independent, the key equivocation can become zero if

$$L \geq H(K)/[\log(\varepsilon) - H(M)], \tag{3.23}$$

in which ε is the size of the alphabet.

Proof

An independent source implies that

$$H(M^L) = L \cdot H(M). \tag{3.24}$$

The total information of the plaintext is equal to L times the information per symbol of the text.

Substitution of eq. (3.24) in eq. (3.21) leads to

$$H(K/C^L) \geq H(K) + L\,[H(M) - \log(\varepsilon)].$$

Thus, the average key equivocation can become zero when

$$L \geq H(K) / [\log(\varepsilon) - H(M)]. \qquad \square$$

Therefore, when the information of the source is low, only a small number of symbols is required to retrieve the key. In practice this can be avoided by ensuring maximum information of the messages, for instance by employing source coding. The value of L for which both sides of (3.23) are equal is called the *unicity distance* (UD). This is the minimum length of the text which is required for finding the key.

Example

Consider an English text which is enciphered by means of a monoalphabetical substitution. Let the information $H(M)$ with respect to the English language be two bits. Since there are 26 letters in the alphabet (disregarding spaces), it follows that:

$$\log(\varepsilon) - H(M) = \log 26 - 2 = 4.7 - 2 = 2.7 \text{ bit.}$$

A monoalphabetical substitution has 26! possibilities for the key. Assuming the uncertainty for each key is the same, the information of each key is equal to:

$$H(K) = \log 26! = 88.38 \text{ bit.}$$

By substituting this value in the expression for the unicity distance, we find:

$$UD = H(K) / [\log(\varepsilon) - H(M)] = 88.38 / 2.7 \approx 32.$$

Therefore, on average it will take a ciphertext of 32 letters to find the correct key.

In the case of a Caesar substitution $H(K)$ decreases to $H(K) = \log 26 = 4.70$ bits. Now, the unicity distance is considerably smaller:

$$UD = 4.70 / 2.7 \approx 2. \qquad\qquad \triangle$$

Finally, we must remark that although the unicity distance gives a value for which $H(K/C^L) = 0$, this does not guarantee that the key can actually be found for every conceivable situation. The statements made here and throughout the entire chapter only apply on average. This is inherent in the use of information theory, which deals with mean values of information, etc. The unicity distance therefore represents the average number of required letters or symbols.

3.4 Error probability and security

In the previous section we use well-defined measures of information, in order to be able to make statements about the absolute security of a system. We saw that the unicity distance gives the minimum length of the ciphertext required for retrieving the key. However, it does not guarantee that the key will actually be found in all cases. This means the error probability is not necessarily equal to zero. We will therefore consider this error probability in more detail.

In this section we will assume the cryptanalyst works logically and takes decisions according to the Bayes model. Suppose the cryptanalyst must make a choice between I hypotheses H_1, H_2, ..., H_ℓ with known corresponding uncertainties $p_i = p(H_i)$, $i = 1,...,\ell$. The conditional probability of a hypothesis H_i, under the condition of a given observation x, is expressed as $P(H_i/x)$. The Bayes rule states that the best choice of hypothesis is the hypothesis with the highest conditional probability, given the occurrence of x. The corresponding error probability Pe is :

$$Pe(H/x) = 1 - E_x [\max P(H_i/x)],$$

in which E_x is the expectation with respect to the variable x.

We will assume that the hypotheses apply to the reconstruction of the plaintext, finding the key, etc., whereas the ciphertext represents an observation. A cryptanalyst can choose between three possibilities with associated error probabilities (note the analogy with the equivocations of Section 3.2):

(i) $Pe(K/C)$ = error probability with respect to finding the key given the ciphertext (ciphertext-only-attack)

(ii) $Pe(M/C)$ = error probability with respect to retrieving the plaintext, based solely on the ciphertext (ciphertext-only-attack)

(iii) $Pe(K/C,M)$ = error probability with respect to finding the key given both the ciphertext and the plaintext (known-plaintext-attack).

The error probability $Pe(K/C)$ is defined as:

$$Pe(K/C) = 1 - E_C \left[\max_K P(M/C) \right].$$

Definitions for $Pe(K/C)$ and $Pe(K/C,M)$ can be made in a similar manner.

We will now consider these three error probabilities as they apply to a transposition cipher. Although a transposition cipher has become a classical cipher system, examination of this cipher is still justified since many modern cryptographic algorithms are constructed from classical systems, e.g. the DES algorithm which is explained in the next chapter.

Let ε denote the number of characters of the alphabet and T denote the period of the key word. The length of the message is given by $L = nT$.

The notion of a *pure cipher*, as introduced by Shannon, plays an important role in the analysis of transposition ciphers by means of error probabilities. If κ represents the set of all transformations, then a cipher system is called a pure cipher if for every transformation t_j, t_k and $t_\ell \in \kappa$, a $t_i \in \kappa$ exists such that

$$t_j\, t_k^{-1}\, t_\ell = t_i,$$

and all keys have the same probability. It is thus clear that a transposition cipher is a pure cipher.

An important characteristic of pure ciphers is that ciphertexts and plaintexts can be divided into a set of residual classes.

In Section 2.2 we saw that the transposition of a message M of length $L = nT$ basically involved no more than exchanging the columns of a matrix. If the message is given by $M^{nT} = [a_1,...,a_{nT}]$, then its transposition is performed by interchanging the columns

$$\underline{a}_i = (a_i, a_{T+i},...,a_{(n-1)T+i})^t.$$

Consider Table 3.1. This table contains all possible plaintexts and ciphertexts with $T = 3$, $L = 6$ and $\varepsilon = 2$, amounting to a total of 64 (i.e. for the general case ε^{nT}).

A residual class consists of all ciphertexts and plaintexts which can be obtained by permutation. Consider the following set: $[(\begin{smallmatrix}000\\001\end{smallmatrix}), (\begin{smallmatrix}000\\010\end{smallmatrix}), (\begin{smallmatrix}000\\10\text{-}0\end{smallmatrix})]$. If a ciphertext is given as $(\begin{smallmatrix}000\\010\end{smallmatrix})$, then the corresponding plaintext must also be an element of this set.

We will use the following notation:

d is the number of elements in a residual class,

Table 3.1. The residual classes of a binary alphabet (period $T = 3$ and the length of the message $L = 6$).

$d = 1$	000 / 000	111 / 000	000 / 111	111 / 111	4 residual classes with 1 element
$d = 3$	000 / 001	001 / 000	000 / 011	011 / 000	
	000 / 010	010 / 000	000 / 101	101 / 000	
	000 / 100	100 / 000	000 / 110	110 / 000	
$d = 3$	001 / 001	110 / 001	001 / 110	111 / 001	12 residual classes with 3 elements
	010 / 010	101 / 010	010 / 101	111 / 010	
	100 / 100	011 / 100	100 / 011	111 / 100	
$d = 3$	001 / 111	011 / 011	111 / 011	011 / 111	
	010 / 111	101 / 101	111 / 101	101 / 111	
	100 / 111	110 / 110	111 / 110	110 / 111	
$d = 6$	001 / 010	011 / 110	011 / 001	001 / 011	4 residual classes with 6 elements
	010 / 001	011 / 101	011 / 010	010 / 011	
	001 / 100	101 / 110	101 / 001	001 / 101	
	010 / 100	101 / 011	101 / 100	100 / 101	
	100 / 010	110 / 011	110 / 010	010 / 110	
	100 / 001	110 / 101	110 / 100	100 / 110	

A_d is the number of residual classes with d elements.

With $T = 3$, as is the case in Table 3.1, d assumes the values 1, 3 and 6. Table 3.2 lists the values of d for other periods.

Table 3.2 also gives the distribution P of the columns within a residual class with d elements. For instance, with $T = 3$ and $d = 3$, the table states that $P = 1^1, 2^1$. This means that with $T = 3$, every message of a residual class with $d = 3$ elements consists of 1 single column and 2 equal columns (see also Table 3.1). $P = 1^3$ applies to messages with 3 different columns and $P = 3^1$ to messages with 3 equal columns, etc. Finally, in Table 3.2, $\Sigma \alpha_m$

Table 3.2. The number of elements d of residual classes as a function of the period T.

T	Σa_m	P	d	T	Σa_m	P	d	T	Σa_m	P	d
1	1	1^1	1	6	1	6^1	1	8	1	8^1	1
2	1	2^1	1		2	$1^1,5^1$	6		2	$1^1,7^1$	8
	2	1^2	2			$2^1,4^1$	15			$2^1,6^1$	28
3	1	3^1	1			3^2	20			$3^1,5^1$	56
	2	$1^1,2^1$	3		3	$1^2,4^1$	30			4^2	70
	3	1^3	6			$1^1,2^1,3^1$	60		3	$1^2,6^1$	56
4	1	4^1	1			2^3	90			$1^1,2^1,5^1$	168
	2	$1^1,3^1$	4		4	$1^3,3^1$	120			$1^1,3^1,4^1$	280
		2^2	6			$1^2,2^2$	180			$2^2,4^1$	420
	3	$1^2,2$	12		5	$1^4,2$	360			$2^1,3^2$	560
	4	1^4	24		6	1^6	720		4	$1^3,5^1$	336
5	1	5^1	1	7	1	7^1	1			$1^2,2^1,4^1$	840
	2	$1^1,4^1$	5		2	$1^1,6^1$	7			$1^2,3^2$	1120
		$2^1,3^1$	10			$2^1,5^1$	21			$1^1,2^2,3^1$	1680
	3	$1^2,3^1$	20			$3^1,4^1$	35			2^4	2520
		$1^1,2^2$	30		3	$1^2,5^1$	42		5	$1^4,4^1$	1680
	4	$1^3,2^1$	60			$1^1,2^1,4^1$	105			$1^3,2^1,3^1$	3360
	5	1^5	120			$1^1,3^2$	140			$1^2,2^3$	5040
						$2^2,3^1$	210		6	$1^5,3^1$	6720
					4	$1^3,4^1$	210			$1^4,2^2$	10080
						$1^2,2^1,3^1$	420		7	$1^6,2^1$	20160
						$1^1,2^3$	630		8	1^8	40320
					5	$1^4,3^1$	840				
						$1^3,2^2$	1260				
					6	$1^5,2$	2520				
					7	1^7	5040				

represents the total number of different columns of a message in a residual class of d elements. It follows that for Table 3.1, $A_1 = 4$, $A_3 = 12$ and $A_6 = 4$. In general, A_d will depend on n and ε (see Table 3.3).

The total number of messages ε^{nT} can easily be calculated and is (see Table 3.1):

$$\varepsilon^{nT} = \sum_d d \cdot A_d. \tag{3.25}$$

The term $\sum_d A_d$ can also be expressed as a function of n, ε and T. As the messages of a residual class can always be found by permutation of the columns of the other messages of that class, the number of residual classes is determined by the total number of different columns, regardless of the order in which they occur.

This effectively results in selecting T columns from a set of ε^n columns. We have already come across this kind of combinatory problem in our discussion of the Hagelin machine (see Section 2.4). In the same manner, we find:

$$\sum_d A_d = \binom{T + \varepsilon^n - 1}{T} \tag{3.26}$$

We are now able to find expressions for the error probabilities as stated in the following theorem.

Theorem 3.5
Assuming all the characters of an alphabet have equal probabilities of occurrence and that all keys have the same uncertainty, the error probabilities can be expressed as:

(a) $$Pe(K/C) = 1 - 1/T! \tag{3.27}$$

(b) $$Pe(M/C) = 1 - \sum_d A_d / \varepsilon^{nT} = 1 - \binom{T + \varepsilon^n - 1}{T} / \varepsilon^{nT} \tag{3.28}$$

(c) $$Pe(K/M,C) = 1 - \sum_d d^2 A_d / (T! \, \varepsilon^{nT}). \tag{3.29}$$

Proof
(a) $$Pe(K/C) = 1 - E_C \left[\max_K P(K/C) \right]. \tag{3.30}$$

Because the ciphertext and the key are mutually independent, it holds that $P(K/C) = P(K)$. With $T!$ possibilities for the key, each with the same probability, the uncertainty of each key is $P(K) = 1/T!$. When this is substituted in eq. (3.30), eq. (3.27) results in

(b) $$Pe(M/C) = 1 - E_C\left[\max_M P(M/C)\right]$$

$$= 1 - \sum_C \left[\max_M P(M/C)\right] P(C), \qquad (3.31)$$

as the sum for all possible C. Remembering that the d elements of a residual class all have the same uncertainty, we can write:

$$\max_M P(M/C) = \frac{1}{d}$$

Furthermore, as the uncertainty of each character is the same, we also find:

$$P(C) = \frac{1}{\varepsilon^{nT}}.$$

This applies for a single ciphertext. However, a residual class contains d elements and there are A_d residual classes, so with dA_d elements in total, $[\max_M P(M/C)] P(C) = 1/d\varepsilon^{nT}$. Therefore, the summation for all d of eq. (3.31) must also be calculated in order to arrive at the following expression for the error probability:

Table 3.3. The number of residual classes A_d with d elements as a function of n and ε ($T = 3$).

	$\varepsilon = 2$			$\varepsilon = 3$		
n	A_1	A_3	A_6	A_1	A_3	A_6
1	2	2	0	3	6	1
2	4	12	4	9	72	84
3	8	56	56	27	702	2 925
4	16	240	560	81	6 480	85 320
5	32	992	4 960	243	58 806	2 362 041
6	64	4 032	41 664	729	530 712	64 304 604
7	128	16 256	341 336	2 187	4 780 782	1 741 001 445
8	256	65 280	2 763 520	6 561	43 040 160	47 050 074 890
9	512	261 632	22 238 720	19 683	387 400 806	1 270 739 210 481
10	1 024	1 047 552	178 433 024	59 049	3 486 725 352	34 313 445 309 924

$$Pe(M/C) = 1 - \sum_d \frac{1}{d\varepsilon^{nT}} \, dA_d = 1 - \sum_d A_d/\varepsilon^{nT}.$$

With eq. (3.26) we can find the desired result.

(c) $$Pe(K/M,C) = 1 - E_{M,C}[\max_K P(K/M,C)]$$

$$= 1 - \sum_{M,C}[\max_K P(K/M,C)] \, P(M,C), \qquad (3.32)$$

summed for all possible pairs of M and C. Within a single residual class, it holds that:

$$\max_K P(K/M,C) = \frac{d}{T!}.$$

The probability of occurrence of a specific pair (M,C) is determined by the probability of M and subsequent d possibilities for the ciphertexts.

$$P(M,C) = \frac{1}{\varepsilon^{nT}} \frac{1}{d}.$$

This probability applies to a single pair (M,C) only. As there are d^2 possible combinations per class and A_d classes with d elements, the resulting error probability can be written as:

$$Pe(K/M,C) = 1 - \sum_d \frac{d}{T!} \cdot \frac{1}{\varepsilon^{nT}} \cdot \frac{1}{d} \, d^2 A_d$$

$$= 1 - \sum_d d^2(A_d/(T! \, \varepsilon^{nT})). \qquad \square$$

Note that we do not (yet) have an explicit expression for $\sum_d d^2 A_d$ in terms of ε, n and T. However, we can provide an upper limit for $Pe(K/M,C)$.

Theorem 3.6

Let $Pe(K/M,C)$ be the error probability with respect to the key given both a ciphertext and a plaintext. Then:

$$Pe(K/M,C) \leq 1 - \varepsilon^{nT}/\left\{ T! \binom{T + \varepsilon^n - 1}{T} \right\}. \qquad (3.33)$$

Proof

Since $(\Sigma ab)^2 \leq \Sigma a^2 \cdot \Sigma b^2$ (Hardy, Littlewood and Polya, 1973) and with $a = d\sqrt{A_d}$ and $b = \sqrt{A_d}$ we find that:

$$\left\{ \sum_d dA_d \right\}^2 \leq \left\{ \sum_d d^2 A_d \right\} \cdot \left\{ \sum_d dA_d \right\}. \tag{3.34}$$

Substitution of (3.25) and (3.26) in (3.34) yields

$$\sum_d d^2 A_d \geq \varepsilon^{2nT} / \left(\frac{T + \varepsilon^n - 1}{T} \right),$$

which can be combined with eq. (3.29) to provide an upper boundary for eq. (3.33). $\quad\square$

Values for the error probabilities of the example shown in Table 3.1 can easily be calculated by entering $T = 3$, $n = 2$ and $\varepsilon = 2$ in eqs. (3.27)–(3.29). This results in:

$$Pe(K/C) = 1 - 1/6 \approx 0.83,$$

$$Pe(M/C) = 1 - 20/64 \approx 0.69,$$

$$Pe(K/M,C) = 1 - 256/384 \approx 0.33.$$

The upper boundary for $Pe(K/M,C)$ is found to be:

$$Pe(K/M,C) \leq 1 - 64/120 \approx 0.53.$$

In Figure 3.4 $Pe(M/C)$ and $Pe(K/C)$ are plotted for several other values of n, ε and T, with the aid of Theorem 3.5. To facilitate the interpretation of these functions, the plots are interpolated between integer values for n even though only integers are permitted for n.

The error probability $Pe(K/C)$ remains constant as a function of n and ε, since the key does not depend on either the number of blocks of length T or the number of characters of the source. Obviously, this does not hold for $Pe(M/C)$.

$Pe(M/C)$ increases as the number of symbols of the source alphabet increases or as the period becomes longer. It also increases when the number of blocks increases. For this last case, if n increases, then $Pe(M/C)$ will approach the curve of $Pe(K/E)$, although it will never exceed it. This is confirmed by the following theorem:

Theorem 3.7

(i) $Pe(M/E) \leq Pe(K/C),$ (3.35)

(ii) $\lim_{n \to \infty} Pe(M/C) = Pe(K/C).$ (3.36)

Proof

(i) $Pe(M/C)$ of eq. (3.28) can be rewritten as:

$$Pe(M/C) = 1 - \binom{T + \varepsilon^n - 1}{T} / \varepsilon^{nT}$$

$$= 1 - \left[\prod_{j=0}^{T-1} (\varepsilon^n + j) \right] / (T! \, \varepsilon^{nT})$$

$$= 1 - \left[\varepsilon^{nT} + \binom{T}{2} \varepsilon^{n(T-1)} + \dots \right] / (T! \, \varepsilon^{nT})$$

$$= 1 - \frac{1}{T!} - \frac{1}{2(T-2)!} \frac{1}{\varepsilon^n} + O(1/\varepsilon^n). \quad (3.37)$$

From this it follows that

$$Pe(M/C) \leq 1 - \frac{1}{T!} \, ,$$

in which the right-hand side of the inequality is precisely equal to $Pe(K/C)$.

(ii) If we take the limit $n \to \infty$ in eq. (3.37) we finally find:

Figure 3.4. Error probabilities of the key and plaintext with $T = 3$, 4 and 5 and with $\varepsilon = 2$ and 3. The plot of $Pe(K/C)$ is indicated by a.

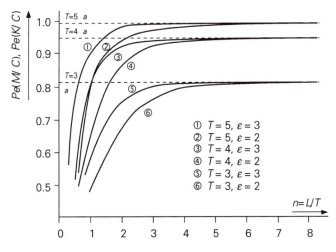

$$\lim_{n\to\infty} Pe(M/C) = 1 - \frac{1}{T!} = Pe(K/C) \qquad\qquad \square$$

From this theorem and Figure 3.4 we can conclude that knowledge of the ciphertext will never lead to a decrease in the error probabilities as n increases.

Figure 3.5 depicts $Pe(K/M,C)$ as a function of n, ε and T as well as, $Pe(K/C)$. It shows that as n increases, $Pe(K/M,C)$ decreases and that the error probability $Pe(K/M,C)$ is always smaller than $Pe(K/C)$. This agrees with the fact that a known-plaintext-attack will always provide more information than a ciphertext-only-attack.

As was the case for information measures, we can again define a sort of unicity distance or, more appropriately, the *error probability distance*, which is the minimum length of the ciphertext required for the error probability $Pe(K/M,C)$ to drop below a given threshold.

The error probability distance is defined as:

Figure 3.5. The error probability $Pe(K/M,C)$ as a function of n ($T = 3$, 4 and 5, $\varepsilon = 2$ and 3). The functions of $Pe(K/C)$ are indicated with an a.

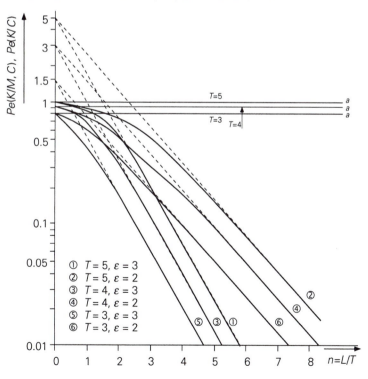

$$PeD(\beta) = \lim_{L} \{L \mid Pe(K/M^L, C^L) \leq \beta\}. \tag{3.38}$$

Without proof, it can be stated that a series expansion of $Pe(K/M,C)$ for a large enough value of n yields:

$$PeD(\beta) \approx T \log_{\varepsilon} \left[\binom{T}{2} / 2\beta \right] \tag{3.39}$$

The strength of the cipher system will increase as $PeD(\beta)$ increases. According to eq. (3.39), the strength of the system is proportional to T.

Normally, before data is enciphered, coding techniques are used to transform the data into a form in which the source symbols have a more-or-less uniform distribution. This process conceals the statistical characteristics of the data and frustrates a ciphertext-only-attack. However, it will still be possible to provide the same analysis of the error probabilities in the case of a non-uniform symbol distribution, although here we will refrain from doing so.

3.5 Practical security

The previous section was mainly concerned with what could be referred to as the theoretical security of a cipher system. We provided an estimate of the security in terms of information measures, error probabilities, etc., based on the statistical properties of the plaintext, ciphertext and the key. These statements were made without paying any attention to the technical capabilities of the cryptanalyst. Suppose a cipher system is based on secret keys. If cryptanalysts are not capable of cracking the system in an 'intelligent' manner, they can always attempt to find the key simply through trial and error. This aspect of security is referred to as practical security and is concerned with what cryptanalysts could achieve given unlimited time and calculating power. This point is becoming ever more important, as technical developments are continually providing bigger and faster computers. It is therefore becoming more and more attractive to try a large number of keys, despite the size of the key space, as this can be done in a relatively short length of time. Table 3.4 shows the time required for finding a key as a function of the number of bits of the key. As is apparent from the table, the required time increases exponentially with the length of the key.

In practice, a system designer must take into account these technical consideration, especially the expectations for the not too distant future, when considering the length of the key.

A more absolute limit has been suggested by Davies and Price (1989). They state an upper limit for the physically achievable calculating power. A binary operation requires a quantity of energy roughly equal to kT, where k is Boltzmann's constant and T the absolute temperature. Under the unrealistic assumption that all the energy of the sun that reaches the earth is used for performing calculations and that the calculations are made at 100 K, it can be demonstrated that approximately 3×10^{45} operations can be performed per year. Davies and Price then contemplated the following with regard to memory capacity. Assume that the storage of 1 bit requires 10 atoms of silicon and a memory device is constructed which is 1000 m high and covers every continent on earth. This memory would be capable of storing no more than 10^{45} bits. Therefore, computations which require more than 10^{46} operations and 10^{45} bits of memory can be regarded as physically impossible. Obviously, calculations requiring far fewer operations and less memory must already be considered totally impractical.

Table 3.4. Required time for finding a key of various lengths.

# bits/key	LSI-chip (10^3 keys/s)	MICRO (10^6 keys/s)	10^6 CHIPS (10^{12} keys/s)
24	2.3 h	8.4 s	—
32	25 d	36 min	2.2 ms
40	18 y	7 d	549 ms
48	4463 y	4,5 y	2.4 min
56	1.1×10^6 y	1.1×10^3 y	10 h
64	2.9×10^8 y	2.9×10^5 y	107 d
128	5.4×10^{27} y	5.4×10^{24} y	5.4×10^{18} y

4

The Data Encryption Standard

4.1 The DES algorithm

The abbrevation DES stands for Data Encryption Standard. This algorithm is used for enciphering blocks of data and offers cryptographic protection of data during storage and transmission.

In 1972 the US Federal Department of Commerce took precautions to improve national security by calling for a cryptographic standard for storing, processing and distributing information, as a result of the increasing number of applications of computer systems. An appeal was made for proposals for a cryptographic algorithm, the specifications of which were defined as follows:

- high level of security;
- comprehensive and transparent specification;
- security may not rely on the secrecy of the algorithm;
- available and accessible to all users;
- suitable for a variety of applications;
- low cost implementation;
- able to be exported;
- accessible for validation.

When no proposals were made, a new call was made in 1974. IBM had been developing a cryptographic algorithm since 1968 and this time IBM responded with a variation on the Lucifer cipher-system. This system enciphers blocks of 128 bits of data with a 128-bit key. After the proposal had been analysed, a great many modifications were made and it was finally published in 1976 under the name DES. A year later the standard was made the Federal Standard in the USA. Although DES now is the most frequently used algorithm for block encipherment, it has never become an international standard and probably never will. This is due not to the quality of DES, but rather to the fact that the ISO (International Organisation for Standardisation) has decided no longer to standardise cryptographic algorithms, only to register them.

We shall first make a small diversion by taking a brief look at the specific problems of standardisation with respect to cryptography. In general, one of the main purposes of standardisation is to facilitate communication between two parties. In the case of communication in a well-defined group (organisation, company etc.) there is no direct need for standardisation as agreements on the use of cryptographic algorithms and communication protocols can be made locally, even after the equipment has been purchased and installed. However, when communication takes place between two different organisations, then some kind of standardisation is indispensable. Consider, for instance, the financial traffic between banks.

An major advantage of standardisation is the fact that before an algorithm or protocol actually becomes a standard, it has already withstood extensive usage: improvements have been made for efficiency; it has been technically optimised; it has been tested for weaknesses etc. The user will therefore have some kind of guarantee of the capabilities of the standard, for instance with respect to security. If after some time the standard still shows some weaknesses, then everyone will experience the same disadvantages, due to the popularity and consequent widespread usage of standards.

Standardisation does, however, have some obvious disadvantages, especially where cryptographic algorithms are concerned. One disadvantage, for instance, is the fact that standards are designed to be used on a large scale, which in itself makes it attractive for hackers to try to crack the system. If a single comprehensive cryptographic algorithm were set as the standard, then all efforts could be focused on cracking this one system. This is the underlying reason for refraining from introducing international standards for cryptographic standards. In addition, standardisation implies publication and therefore the security of the system will depend entirely on the strength of the algorithms and protocols themselves. The algorithms and protocols would therefore need to meet stricter requirements.

Finally, standardisation of products will result in a growth of the market and, consequently, lower prices. People will be more willing to buy cryptographic equipment and the number of different applications will increase. However, this increase in use obviously results in a deterioration of the national security.

We will now return to DES. DES assumes the data are available in binary form. It is designed for enciphering and deciphering blocks of data of 64 bits. The key is also 64 bits long, of which in fact only 56 bits are used. The remaining 8 bits are used for parity checks. The total number of keys is thus $2^{56} = 7.2 \times 10^{16}$.

The DES algorithm essentially consists of a series of permutations and substitutions. A block which is to be enciphered is first subjected to an initial permutation IP, then to a complex series of key-dependent operations and finally to a permutation IP^{-1}, which is the inverse of the initial permutation.

The DES algorithm has also been implemented in an LSI chip, which is now readily available. This is particularly suitable for high speed applications (400 Kbits/s – 50 Mbits/s) and real-time encipherment, for instance for encrypting speech.

Encipherment and decipherment

In this chapter we will employ the following notation. Assume two blocks of data of L and R bits are given (here, L and R stand for left and right, respectively) and that LR denotes the block consisting of the bits of L followed by the those of R. Since the DES algorithm operates on blocks of 64 bits, L represents the 32 left-hand bits and R represents the 32 right-hand bits.

Figure 4.1 shows the general diagram of the DES algorithm. The input consists of 64 bits, which are first subjected to an initial permutation. After this permutation, the 64 bits are split into 32 left-hand bits and 32 right-hand bits. Subseqently, 16 operations, the so-called *DES rounds*, are performed, after which the inverse permutation of the blocks of data is calculated, resulting in an enciphered output.

Considering the 16 operations of the algorithm, we see that for every step the 32 left-hand bits are immediately exchanged for the 32 right-hand bits, the original 32 left-hand bits being first processed in some manner, before taking the place of the right-hand bits. The processing consists of a binary addition of the original 32 left-hand bits and a given function $F(\cdot,\cdot)$, which depends on the 32 right-hand bits and a given key. K_1–K_{16} are DES subkeys which are derived from the 64 bits of the original key K. Later in this section exactly how the 16 subkeys are calculated from the original key and how the value of the function $F(\cdot,\cdot)$ can be determined will be explained.

It is evident that for each of the 16 steps, the following holds: assuming the output of each step is denoted by $L'R'$ and the input is given by LR, we can write the relations between these two as:

$$L' = R,$$

$$R' = L + F(R,K),$$

in which + represents a binary addition and K is a subkey.

If the permutated input (i.e. the input immediately after the initial permutation IP) is denoted by L_0R_0 by and the pre-output (i.e. the output

Figure 4.1. The general diagram of the DES algorithm.

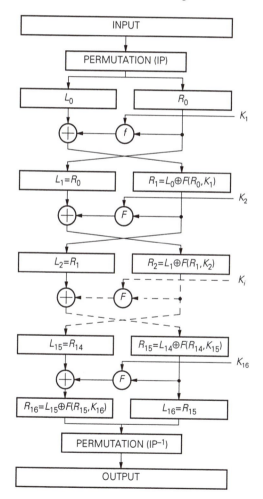

directly before the inverse IP^{-1} permutation) by $L_{16}R_{16}$, then the following expressions are found for the intermediate levels:

$$L_n = R_{n-1}, \tag{4.1}$$

$$R_n = L_{n-1} + F(R_{n-1},K_n), \tag{4.2}$$

$$K_n = G(K,n). \tag{4.3}$$

The last function $G(\cdot,\cdot)$ indicates that the subkey K_n is a function of the original key K of 64 bits and the number n of the processing step.

The above text describes the method of encipherment. However, decipherment is performed in precisely the same manner. The DES algorithm is constructed in such a way that the same routines can be used for both encipherment and decipherment. Now $R_{16}L_{16}$ forms the permutated input. The same steps are made as previously, with the only difference being the order of the subkeys, which is now reversed. Thus, K_{16} is used first, then K_{15} etc. Encipherment then occurs according to:

$$R_{n-1} = L_n, \tag{4.4}$$

$$L_{n-1} = R_n + F(L_n,K_n), \tag{4.5}$$

$$K_n = G(K,n). \tag{4.6}$$

This means that encipherment and decipherment do not require different chips and that both tasks can be performed using the same IC.

The initial permutation

Before the subsequent operations are performed, each block of 64 bits is subjected to an initial permutation IP. The associated matrix is given by:

IP:

58	50	42	34	26	18	10	2
60	52	44	36	28	20	12	4
62	54	46	38	30	22	14	6
64	56	48	40	32	24	16	8
57	49	41	33	25	17	9	1
59	51	43	35	27	19	11	3
61	53	45	37	29	21	13	5
63	55	47	39	31	23	15	7

This matrix should be interpreted as follows. The 58th bit of the original input is placed at the 1st position, the 50th bit of the original 64 bits is put in 2nd position, the 42nd bit in the 3rd position, etc.

After the 16 operations, the pre-output is permutated according to IP^{-1}, which is given by:

$$
\mathrm{IP}^{-1}:\quad
\begin{array}{cccccccc}
40 & 8 & 48 & 16 & 56 & 24 & 64 & 32 \\
39 & 7 & 47 & 15 & 55 & 23 & 63 & 31 \\
38 & 6 & 46 & 14 & 54 & 22 & 62 & 30 \\
37 & 5 & 45 & 13 & 53 & 21 & 61 & 29 \\
36 & 4 & 44 & 12 & 52 & 20 & 60 & 28 \\
35 & 3 & 43 & 11 & 51 & 19 & 59 & 27 \\
34 & 2 & 42 & 10 & 50 & 18 & 58 & 26 \\
33 & 1 & 41 & 9 & 49 & 17 & 57 & 25
\end{array}
$$

The 40th bit is placed in the 1st position, the 8th bit in the 2nd position etc. It can easily be verified without the intermediate processing steps that the original input is once again obtained.

The cipher function F(R,K)

We saw in the general diagram of the DES algorithm that for each step the 32 right-hand bits are determined by the 32 left-hand bits of the previous step and a given function, the so-called *cipher function*. Figure 4.2 shows this process in more detail.

First a so-called expansion operation E is performed, which expands the sequence of 32 bits to a total of 48 bits. This *expansion operation* can be described by a matrix as follows:

$$
E:\quad
\begin{array}{cccccc}
32 & 1 & 2 & 3 & 4 & 5 \\
4 & 5 & 6 & 7 & 8 & 9 \\
8 & 9 & 10 & 11 & 12 & 13 \\
12 & 13 & 14 & 15 & 16 & 17 \\
16 & 17 & 18 & 19 & 20 & 21 \\
20 & 21 & 22 & 23 & 24 & 25 \\
24 & 25 & 26 & 27 & 28 & 29 \\
28 & 29 & 30 & 31 & 32 & 1
\end{array}
$$

This implies that the 32nd bit of the 32 right-hand bit is placed at the 1st position, the 1st bit in position 2, etc. The final result of the expansion operation is a bit stream of 48 bits, since some bits are used more than once. Figure 4.3 illustrates the mechanism of the expansion operation.

The result of the expansion operation is added to the subkey K, which also consists of 48 bits, as depicted in Figure 4.2. This subkey is derived from the original key K; exactly how this is done will be described later; for the moment it will suffice to state that the 48 bits resulting from the expansion operation are added to the subkey in binary form. This result, in turn, functions as the input of eight so-called *substitution boxes* (*S-boxes*), as is shown in Figure 4.2.

Figure 4.2. The cipher function $F(R,K)$.

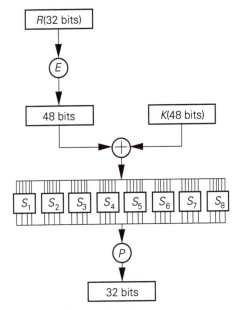

Figure 4.3. The expansion operation E.

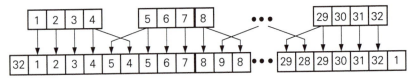

Table 4.1. Table for S-box 4.

	0	1	2	3	4	5	6	7	8	9	10	11	12	13	14	15
0	7	13	14	3	0	6	9	10	1	2	8	5	11	12	4	15
1	13	8	11	5	6	15	0	3	4	7	2	12	1	10	14	9
2	10	6	9	0	12	11	7	13	15	1	3	14	5	2	8	4
3	3	15	0	6	10	1	13	8	9	4	5	11	12	7	2	14

The *S*-boxes are the heart of the DES algorithm. Each *S*-box transforms six input bits to four output bits. Table 4.1 illustrates exactly how this takes place.

Assume the six input bits of the *S*-box are denoted by:

$$B = (x_1, x_2, x_3, x_4, x_5, x_6).$$

The output of the *S*-box is computed as follows. The first and the last bits, x_1 and x_6, determine which row of Table 4.1 is selected, whereas the middle four bits, x_2, x_3, x_4 and x_5, determine which column is used. The output of the *S*-box is the binary representation of the value located at the intersection of the selected row and column in the table.

Example
Let the input of the *S*-box be given by $B = (111011)$. The row is selected by the combination of the first and the last bits: $11 = 3$. The column is selected by the four remaining bits: $1101 = 13$. At the intersection of row 3 and column 13 the table contains the value 7. Hence, the output of the *S*-box is 0111.   △

Table 4.2 provides the tables for all *S*-boxes. The final result of all eight *S*-boxes is a bit stream of 32 bits, which is subsequently subjected to a permutation *P* according to the following table.

	16	7	20	21
	29	12	28	17
	1	15	23	26
P:	5	18	31	10
	2	8	24	14
	32	27	3	9
	19	13	30	6
	22	11	4	25

After all these operations a bit stream of 32 bits is obtained, which is denoted by $F(R,K)$. Returning to Figure 4.1, we see that the binary addition of the 32 left-hand bits and $F(R,K)$ finally produces the 32 right-hand bits of each processing step.

Table 4.2. Tables for all eight *S*-boxes.

S-box 1:

	0	1	2	3	4	5	6	7	8	9	10	11	12	13	14	15
0	14	4	13	1	2	15	11	8	3	10	6	12	5	9	0	7
1	0	15	7	4	14	2	13	1	10	6	12	11	9	5	3	8
2	4	1	14	8	13	6	2	11	15	12	9	7	3	10	5	0
3	15	12	8	2	4	9	1	7	5	11	3	14	10	0	6	13

S-box 2:

	0	1	2	3	4	5	6	7	8	9	10	11	12	13	14	15
0	15	1	8	14	6	11	3	4	9	7	2	13	12	0	5	10
1	3	13	4	7	15	2	8	14	12	0	1	10	6	9	11	5
2	0	14	7	11	10	4	13	1	5	8	12	6	9	3	2	15
3	13	8	10	1	3	15	4	2	11	6	7	12	0	5	14	9

S-box 3:

	0	1	2	3	4	5	6	7	8	9	10	11	12	13	14	15
0	10	0	9	14	6	3	15	5	1	13	12	7	11	4	2	8
1	13	7	0	9	3	4	6	10	2	8	5	14	12	11	15	1
2	13	6	4	9	8	15	3	0	11	1	2	12	5	10	14	7
3	1	10	13	0	6	9	8	7	4	15	14	3	11	5	2	12

S-box 4:

	0	1	2	3	4	5	6	7	8	9	10	11	12	13	14	15
0	7	13	14	3	0	6	9	10	1	2	8	5	11	12	4	15
1	13	8	11	5	6	15	0	3	4	7	2	12	1	10	14	9
2	10	6	9	0	12	11	7	13	15	1	3	14	5	2	8	4
3	3	15	0	6	10	1	13	8	9	4	5	11	12	7	2	14

S-box 5:

	0	1	2	3	4	5	6	7	8	9	10	11	12	13	14	15
0	2	12	4	1	7	10	11	6	8	5	3	15	13	0	14	9
1	14	11	2	12	4	7	13	1	5	0	15	10	3	9	8	6
2	4	2	1	11	10	13	7	8	15	9	12	5	6	3	0	14
3	11	8	12	7	1	14	2	13	6	15	0	9	10	4	5	3

S-box 6:

	0	1	2	3	4	5	6	7	8	9	10	11	12	13	14	15
0	12	1	10	15	9	2	6	8	0	13	3	4	14	7	5	11
1	10	15	4	2	7	12	9	5	6	1	13	14	0	11	3	8
2	9	14	15	5	2	8	12	3	7	0	4	10	1	13	11	6
3	4	3	2	12	9	5	15	10	11	14	1	7	6	0	8	13

S-box 7:

	0	1	2	3	4	5	6	7	8	9	10	11	12	13	14	15
0	4	11	2	14	15	0	8	13	3	12	9	7	5	10	6	1
1	13	0	11	7	4	9	1	10	14	3	5	12	2	15	8	6
2	1	4	11	13	12	3	7	14	10	15	6	8	0	5	9	2
3	6	11	13	8	1	4	10	7	9	5	0	15	14	2	3	12

S-box 8:

	0	1	2	3	4	5	6	7	8	9	10	11	12	13	14	15
0	13	2	8	4	6	15	11	1	10	9	3	14	5	0	12	7
1	1	15	13	8	10	3	7	4	12	5	6	11	0	14	9	2
2	7	11	4	1	9	12	14	2	0	6	10	13	15	3	5	8
3	2	1	14	7	4	10	8	13	15	12	9	0	3	5	6	11

Determination of the subkeys

The previous section described the ins and outs of the DES algorithm. All that remains to be explained is the generation of the subkeys from the main key K. We have already mentioned that, in fact, only 56 bits of the 64 bits of the key are used and that the other 8 bits are parity bits. Figure 4.4 gives the scheme for generating the 16 subkeys from the 64-bit key K. First, the original 64 bits are reduced to 56 bits by PC-1 and the following table:

PC-1:

57	49	41	33	25	17	9
1	58	50	42	34	26	18
10	2	59	51	43	35	27
19	11	3	60	52	44	36
63	55	47	39	31	23	15
7	62	54	46	38	30	22
14	6	61	53	45	37	29
21	13	5	28	20	12	4

Figure 4.4. Computation of the subkeys.

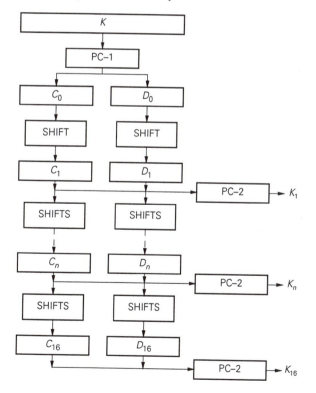

The 57th bit is put at position 1, the 49th bit at position 2 etc. Note that the bits 8, 16, 24 etc. of K are not used.

After PC-1, the 56 bits are split into two groups of 28 bits; a left group (C) and a right group (D). In the following round, C and D are subjected to either 1 or 2 left-shifts. This means that the sequence (1, 2, ..., 28) is transformed to (2, 3, 4, ..., 28, 1) with a single shift, or into (3, 4, 5, ..., 28, 1, 2) after two left-shifts.

The number of shifts depends of the number of the round which is being executed, according to the following table:

round	1	2	3	4	5	6	7	8	9	10	11	12	13	14	15	16
shifts	1	1	2	2	2	2	2	2	1	2	2	2	2	2	2	1

Finally, PC-2 converts the two groups of 28 bits to a single bit stream of 48 bits:

$$
\begin{array}{cccccc}
14 & 17 & 11 & 24 & 1 & 5 \\
3 & 28 & 15 & 6 & 21 & 19 \\
23 & 19 & 12 & 4 & 26 & 8 \\
16 & 7 & 27 & 20 & 13 & 2 \\
41 & 52 & 31 & 37 & 47 & 55 \\
30 & 40 & 51 & 45 & 33 & 48 \\
44 & 49 & 39 & 56 & 34 & 53 \\
46 & 42 & 50 & 36 & 29 & 32 \\
\end{array}
$$

PC-2:

Example

This process is illustrated by a numerical example of a single processing step. For reasons of simplicity, we will assume that the key is identical to the 64 input bits, so: input = K.

Suppose:

input = K = 0 1 2 3 4 5 6 7 8 9 A B C D E F (hexadecimal),

or

input = K = 0000 0001 0010 0011 0100 0101 0110 0111

1000 1001 1010 1011 1100 1101 1110 1111 (binary).

Computation of the subkey K_1 yields:

K_1 =   0000 1011 0000 0010 0110 0111

1001 1011 0100 1001 1010 0101

The initial permutation IP results in:

$$L_0 = 1100\ 1100\ 0000\ 0000\ 1100\ 1100\ 1111\ 1111$$

$$R_0 = 1111\ 0000\ 1010\ 1010\ 1111\ 0000\ 1010\ 1010$$

Continuing with the expansion operation E on R_0, we find:

$$E(R_0) =\ 011110\ 100001\ 010101\ 010101$$

$$011110\ 100001\ 010101\ 010101$$

Binary addition of $E(R_0)$ and K_1 produces:

$$E(R_0) + K_1 =\ 011100\ 010001\ 011100\ 110010$$

$$111000\ 010101\ 110011\ 110000$$

This bit stream is then presented to the eight S-boxes:

S-box no.	input	column	row	value	output
1	011100	14	0	0	0000
2	010001	8	1	12	1100
3	011100	14	0	2	0010
4	110010	9	2	1	0001
5	111000	12	2	6	0110
6	010101	10	1	13	1101
7	110011	9	3	5	0101
8	110000	8	2	0	0000

After executing the permutation P, $F(R_0,K_1)$ is given as:

$$F(R_0,K_1) = 1001\ 0010\ 0001\ 1100\ 0010\ 0000\ 1001\ 1100$$

Adding L_0 to this result leads to:

$$R_1 = L_0 + F(R_0,K_1) = 0101\ 1110\ 0001\ 1100\ 1110\ 1100\ 0110\ 0011$$

Therefore, we finally find for the output of the first processing step that:

$$L_1 R_1 =\ 1111\ 0000\ 1010\ 1010\ 1111\ 0000\ 1010\ 1010$$

$$0101\ 1110\ 0001\ 1100\ 1110\ 1100\ 0110\ 0011.$$

The results of the following 15 steps can be calculated in precisely the same manner. However, here, we will refrain from doing so!   △

4.2 Characteristics of the DES

In this section we will pay attention to a number of characteristics of the DES.

When designing a block cipher-system it is generally desirable to ensure that all output bits depend on all of the input bits. If this is not the case and some output bits depend merely on a small number of input bits, then there will be an increased chance of finding the input bits which correspond to these output bits when a ciphertext-only-attack is attempted. The cryptographic algorithm would therefore be weaker than necessary. However, this is not the case for the DES, as all output bits depend on all input bits. Consider Table 4.3. This table represents the dependence of the 64 output bits on the 64 input bits for the DES as a function of the number of rounds or processing steps. It shows that after 5 rounds the 64 bits of the intermediate results already depend on all 64 input bits.

Another desirable feature is the so-called *avalanche effect*. This effect causes a small alteration of the plaintext to result in a large change of the ciphertext. Suppose that this did not occur and a small alteration of the plaintext results in a nearly identical ciphertext. Let $M1$ denote a plaintext, $C1$ the corresponding ciphertext and $C2$ a ciphertext with unknown plaintext. If the differences between $C1$ and $C2$ are minimal, then $M2$, which we are trying to discover, will differ only marginally from $M1$, since small differences between ciphertexts are a result of small differences between plaintexts. Therefore, by repeatedly altering $M1$ slightly, we can generate a set of plaintexts, which must contain $M2$.

Table 4.4 contains two almost identical plaintexts, which differ in only one bit. The results of each consecutive round are given as well as the differences between these results. The value of δ is the Hamming distance, which represents the number of different bits between the two sequences of 64 bits. We can conclude that after 16 rounds, 34 bits of the sequences are different, which corresponds well to the expected mean difference of 32 bits.

Table 4.5 demonstrates the avalanche effect for two keys with only a one-bit difference, for the same plaintext.

Table 4.3. Interdependence of input and output for the DES.

# bit	round															
	1	2	3	4	5	6	7	8	9	10	11	12	13	14	15	16
0	1	7	34	60	64	64	64	64	64	64	64	64	64	64	64	64
1	1	7	28	54	64	64	64	64	64	64	64	64	64	64	64	64
2	1	7	28	54	64	64	64	64	64	64	64	64	64	64	64	64
3	1	7	30	56	64	64	64	64	64	64	64	64	64	64	64	64
4	1	7	32	58	64	64	64	64	64	64	64	64	64	64	64	64
5	1	7	32	58	64	64	64	64	64	64	64	64	64	64	64	64
6	1	7	31	57	64	64	64	64	64	64	64	64	64	64	64	64
7	1	7	28	54	64	64	64	64	64	64	64	64	64	64	64	64
8	1	7	28	54	64	64	64	64	64	64	64	64	64	64	64	64
9	1	7	32	58	64	64	64	64	64	64	64	64	64	64	64	64
10	1	7	30	56	64	64	64	64	64	64	64	64	64	64	64	64
11	1	7	32	58	64	64	64	64	64	64	64	64	64	64	64	64
12	1	7	32	58	64	64	64	64	64	64	64	64	64	64	64	64
11	1	7	28	54	64	64	64	64	64	64	64	64	64	64	64	64
14	1	7	34	60	64	64	64	64	64	64	64	64	64	64	64	64
15	1	7	30	56	64	64	64	64	64	64	64	64	64	64	64	64
16	1	7	32	58	64	64	64	64	64	64	64	64	64	64	64	64
17	1	7	32	58	64	64	64	64	64	64	64	64	64	64	64	64
18	1	7	32	58	64	64	64	64	64	64	64	64	64	64	64	64
19	1	7	30	56	64	64	64	64	64	64	64	64	64	64	64	64
20	1	7	30	56	64	64	64	64	64	64	64	64	64	64	64	64
21	1	7	30	56	64	64	64	64	64	64	64	64	64	64	64	64
22	1	7	34	60	64	64	64	64	64	64	64	64	64	64	64	64
23	1	7	28	54	64	64	64	64	64	64	64	64	64	64	64	64
24	1	7	30	56	64	64	64	64	64	64	64	64	64	64	64	64
25	1	7	34	60	64	64	64	64	64	64	64	64	64	64	64	64
26	1	7	29	55	64	64	64	64	64	64	64	64	64	64	64	64
27	1	7	32	58	64	64	64	64	64	64	64	64	64	64	64	64
28	1	7	32	58	64	64	64	64	64	64	64	64	64	64	64	64
29	1	7	32	58	64	64	64	64	64	64	64	64	64	64	64	64
30	1	7	33	59	64	64	64	64	64	64	64	64	64	64	64	64
31	1	7	28	54	64	64	64	64	64	64	64	64	64	64	64	64
32	7	34	60	64	64	64	64	64	64	64	64	64	64	64	64	64
33	7	28	54	64	64	64	64	64	64	64	64	64	64	64	64	64
34	7	28	54	64	64	64	64	64	64	64	64	64	64	64	64	64
35	7	30	56	64	64	64	64	64	64	64	64	64	64	64	64	64
36	7	32	58	64	64	64	64	64	64	64	64	64	64	64	64	64
37	7	32	58	64	64	64	64	64	64	64	64	64	64	64	64	64
38	7	31	57	64	64	64	64	64	64	64	64	64	64	64	64	64
39	7	28	54	64	64	64	64	64	64	64	64	64	64	64	64	64
40	7	28	54	64	64	64	64	64	64	64	64	64	64	64	64	64
41	7	32	58	64	64	64	64	64	64	64	64	64	64	64	64	64
42	7	30	56	64	64	64	64	64	64	64	64	64	64	64	64	64
43	7	32	58	64	64	64	64	64	64	64	64	64	64	64	64	64
44	7	32	58	64	64	64	64	64	64	64	64	64	64	64	64	64
45	7	28	54	64	64	64	64	64	64	64	64	64	64	64	64	64
46	7	34	60	64	64	64	64	64	64	64	64	64	64	64	64	64
47	7	30	56	64	64	64	64	64	64	64	64	64	64	64	64	64
48	7	32	58	64	64	64	64	64	64	64	64	64	64	64	64	64
49	7	32	58	64	64	64	64	64	64	64	64	64	64	64	64	64
50	7	32	58	64	64	64	64	64	64	64	64	64	64	64	64	64
51	7	30	56	64	64	64	64	64	64	64	64	64	64	64	64	64
52	7	30	56	64	64	64	64	64	64	64	64	64	64	64	64	64
53	7	30	56	64	64	64	64	64	64	64	64	64	64	64	64	64
54	7	34	60	64	64	64	64	64	64	64	64	64	64	64	64	64
55	7	28	54	64	64	64	64	64	64	64	64	64	64	64	64	64
56	7	30	56	64	64	64	64	64	64	64	64	64	64	64	64	64
57	7	34	60	64	64	64	64	64	64	64	64	64	64	64	64	64
58	7	29	55	64	64	64	64	64	64	64	64	(,4	64	64	64	64
59	7	32	58	64	64	64	64	64	64	64	64	64	64	64	64	64
60	7	32	58	64	64	64	64	64	64	64	64	64	64	64	64	64
61	7	32	58	64	64	64	64	64	64	64	64	64	64	64	64	64
62	7	33	59	64	64	64	64	64	64	64	64	64	64	64	64	64
63	7	28	54	64	64	64	64	64	64	64	64	64	64	64	64	64

There are 2^{64} possibilities for the plaintexts which can be used as the input to the DES algorithm. The number of ciphertexts is also 2^{64}. Each of the ciphertexts must correspond to one of the plaintexts. Therefore, there must exists a value for n, such that:

$$E^n(M) = M, \tag{4.7}$$

by which we mean that if we encipher M n times, we will arrive at M once again. Thus, if a given ciphertext is repeatedly enciphered, at a certain stage the corresponding plaintext is obtained, without actually knowing the key.

Table 4.4. Avalanche effect for the DES: modifications of the plaintext.

round									δ
0	00000000	00000000	00000000	00000000	00000000	00000000	00000000	00000000	
	10000000	00000000	00000000	00000000	00000000	00000000	00000000	00000000	1
1	00000000	00000000	00000000	00000000	10000101	01111110	00101010	01000011	
	00000001	00000000	00000000	00000000	11000001	01111111	00101011	01010011	6
2	10000101	01111110	00101010	01000011	11010111	00101111	00001101	01111011	
	11000001	01111111	00101011	01010011	00011111	11010001	00100001	11011001	21
3	11010111	00101111	00001101	01111011	11000111	01101110	01101100	10110011	
	00011111	11010001	00100001	11011001	01001010	10010100	11010111	11101001	35
4	11000111	01101110	01101100	10110001	01001100	10110000	01110111	10001010	
	01001010	10010100	11010111	11101001	10100000	00011101	10101010	00101111	39
5	01001100	10110000	01110111	10001010	01110010	00101011	10111100	10000001	
	10100000	00011101	10101010	00101111	11111100	01100010	01111110	10010110	34
6	01110010	00101011	10111100	10000001	01011001	10000101	01110010	01111011	
	11111100	01100010	01111110	10010110	11000010	00011100	10001110	01010001	32
7	01011001	10000101	01110010	01111011	10000010	01100111	10101110	10011100	
	11000010	00011100	10001110	01010001	10110100	01011101	10011110	10110000	31
8	10000010	01100111	10101110	10011100	11100111	11011101	11011011	10010100	
	10110100	01011101	10011110	10110000	00100110	10010010	00101000	00010101	29
9	11100111	11011101	11011011	10010100	01110001	10010000	00001111	00010001	
	00100110	10010010	00101000	00010101	00001111	01101011	10110010	10101110	42
10	01110001	10010000	00001111	00010001	00001010	10101101	00110011	11100100	
	00001111	01101011	10110010	10101110	11001100	10000110	00001001	10011111	44
11	00001010	10101101	00110011	11100100	01010001	01100001	10110010	10000001	
	11001100	10000110	00001001	10011111	11110000	00000110	01110111	10010000	32
12	01010001	01100001	10110010	10000001	01111101	11011101	01001010	10011110	
	11110000	00000110	01110111	10010000	00101000	11111110	01110101	11111010	30
13	01111101	11011101	01001010	10011110	01110101	00010111	00111001	00101000	
	00101000	11111110	01110101	11111010	11011001	00000010	10111010	11100100	30
14	01110101	00010111	00111001	00101000	10011101	10100000	00011110	01001110	
	11011001	00000010	10111010	11100100	00111111	00100110	11010111	00001111	26
15	10011101	10100000	00011110	01001110	10111011	00010100	11111100	11110010	
	00111111	00100110	11010111	00001111	01010010	00011101	01000001	00011010	29
16	11000100	11010111	00101100	10011101	11101110	11011110	01011110	10001011	
	00101100	10010111	01100000	01110110	10100111	00000101	10001101	01000100	34
K	00000011	10010110	01001001	11000101	00111001	00110001	00111001	01100101	

Luckily though, this is not a serious weakness of the DES, since n will be of an order of magnitude which renders this approach as unattractive as trying all possible keys. The only exception to this statement is when one of the four weak keys of the DES is used. It is characteristic of these weak keys that when a plaintext is enciphered twice with one of them, the original plaintext will result. So therefore, $n = 2$. These *weak keys* are given in hexadecimal representation by:

Table 4.5. Avalanche effect for the DES: alterations to the key.

round									δ
0	01101000	10000101	00101111	01111010	00010011	01110110	11101011	10100100	
	01101000	10000101	00101111	01111010	00010011	01110110	11101011	10100100	0
1	11000010	11101101	01001101	01111100	00111101	01001000	10110010	00101000	
	11000010	11101101	01001101	01111100	00111101	00001000	10100010	00101000	2
2	00111101	01001000	10110010	00101000	00110111	01011110	01101101	11111000	
	00111101	00001000	10100010	00101000	01100010	01110111	00001100	01111100	14
3	00110111	01011110	01101101	11111000	11111000	01000100	11111010	00010100	
	01100010	01110111	00001100	01111100	10100011	01110111	00110010	01100110	28
4	11111000	01000100	11111010	00010100	01011110	01010101	00001000	01101110	
	10100011	01110111	00110010	01100110	11001011	11011001	01011001	00011101	32
5	01011110	01010101	00000000	01101110	11010111	10111111	11010111	00010110	
	11001011	11011001	01011001	00011101	10101101	11111111	00011011	01011000	30
6	11010111	10111111	11010111	00010110	11111001	11101000	10010011	01000111	
	10101101	11111111	00011011	01011000	00100101	10000110	01010110	00100010	32
7	11111001	11101000	10010011	01000111	00100001	00110100	01110110	10110110	
	00100101	10000110	01010110	00100010	11101011	11011101	00010111	10011001	35
8	00100001	00110100	01110110	10110110	10010000	10110010	01011001	11101110	
	11101011	11011101	00010111	10011001	01100010	10100100	11110100	11010010	34
9	10010000	10110010	01011001	11101110	10011011	01011010	10110110	01000101	
	01100010	10100100	11110100	11010010	01110100	11101100	11101000	10101011	40
10	10011011	01011010	10110110	01000101	10001010	11111110	10010001	10000010	
	01110100	11101100	11101000	10101011	10010101	01001011	11110001	00100000	38
11	10001010	11111110	10010001	10000010	10111110	01000110	10011011	00011111	
	10010101	01001011	11110001	00100000	01000010	01101111	00110001	10001110	31
12	10111110	01000110	10011011	00011111	11111011	10010110	00100100	01110010	
	01000010	01101111	00110001	10001110	11001111	01001011	11010010	00110011	33
13	11111011	10010110	00100100	01110010	01011111	11110111	11010011	00110000	
	11001111	01001011	11010010	00110011	00011011	10110011	01010110	10010110	28
14	01011111	11110111	11010011	00110000	10110001	00010100	01101011	01010101	
	00011011	10110011	01010110	10010110	01111111	10010110	11100000	10011100	26
15	10110001	00010100	01101011	01010101	10011101	01110010	00010011	00010010	
	01111111	10010110	11100000	10011100	00100110	00111111	01000100	00100111	34
16	11001000	01101100	10010000	10010001	10101011	01110001	01100101	10000001	
	00100011	11100010	11101011	00100100	00110101	10110010	01011100	00010001	35
K	11100100	11110111	11011110	00110001	00111011	00001000	01100011	11011100	
K'	01100100	11110111	11011110	00110001	00111011	00001000	01100011	11011100	

01	01	01	01	01	01	01	01,
FE	FE	FE	FE	FE	FE	FE	FE,
1F	1F	1F	1F	0E	0E	0E	0E,
E0	E0	E0	E0	F1	F1	F1	F1.

In practice, care should be taken to avoid the use of these keys.

In addition, we must also mention the existence of *semi-weak keys*. With these a plaintext is enciphered first with one key and subsequently with another, different key, the same plaintext is once again obtained. There are six pairs of these semi-weak keys:

01	FE	01	FE	01	FE	01	FE
FE	01	FE	01	FE	01	FE	01

1F	E0	1F	E0	0E	F1	0E	F1
E0	1F	E0	1F	F1	0E	F1	0E

01	E0	01	E0	01	F1	01	F1
E0	01	E0	01	F1	01	F1	01

1F	FE	1F	FE	0E	FE	0E	FE
FE	1F	FE	1F	FE	0E	FE	0E

01	1F	01	1F	01	0E	01	0E
1F	01	1F	01	0E	01	0E	01

E0	FE	E0	FE	F1	FE	F1	FE
FE	E0	FE	E0	FE	F1	FE	F1

These semi-weak keys must also be avoided in practice.

A surprising property of the DES is that:

$$\text{if } E_K(M) = C, \text{ then } E_{\overline{K}}(\overline{M}) = \overline{C}, \tag{4.8}$$

where the bar over the symbols indicates that these are the complements of K, M and C.

This property enables an exhaustive key search to be reduced to half its length: 2^{55} instead of 2^{56} possibilities. Let $C1 = E_K(M)$ and $C2 = E_K(\overline{M})$, then due to the given property we immediately find: $\overline{C2} = E_{\overline{K}}(M)$. We can therefore conclude that only half the number of keys need be tested. Encipher M with all 2^{55} possible keys with the last bit equal to zero; if this results in $C1$, then K is found, otherwise, if the result is the complement of $C2$, then the key is given by the complement of K: $\overline{K}$.

No methods have yet been developed which can lead to the key faster than a exhaustive search in the key space of 2^{55} elements. Biham and Shamir (1991) undertook an attack based on differential cryptanalysis. This technique concentrates on an analysis of the effect of small differences between two plaintexts and the resulting differences in the ciphertexts. These differences can be used to ascribe certain probabilities to possible keys and to determine the most probable key. The complexity depends on the number of rounds and is given in Table 4.6. For the DES and DES-like algorithms with less than 15 rounds this method provides a considerable reduction of the complexity. However, surprisingly, for 16 rounds, which is exactly the case for the DES, this method results in an increase in complexity, compared to the previously found value of 2^{55}. So luckily, this method is no serious threat for the DES itself, only for DES-like algorithms of less than 16 rounds. As far as the DES is concerned, the quickest method is still to test all 2^{55} keys.

In Section 3.5 we examined practical security. Apparently, if we had 10^6 chips, which could each test 10^6 keys of 56 bits/s, the average time for finding a key would be about 10 hours. Does this form a serious threat for the DES? Diffie and Hellman (1977a) made some calculations of the actual costs for finding a key, given a time span of half a day. They assumed a price of $10 per chip and twice this amount for the design and control-hardware (obviously, somehow one must keep track of which chip has tried which key). These assumptions lead to a total cost of $20 million. If this value is written off over a period of 5 years, each solution costs $5000. Thus, considerable investment is required to find the keys of the DES with an exhaustive search. In 1989 IBM's estimate of these costs turned out higher, i.e. $200 million. In 1993 Bell-Northern estimated that a total investment of $100 million would be needed to find the key within 20 minutes, assuming that 3×10^{13} keys could be tried every second.

4.3 Alternative descriptions

The actual design criteria of the DES algorithm are kept secret by the designers. The functioning of the algorithm is only described by given

Table 4.6. Complexity of the Bihar–Shamir attack as a function of the number of rounds.

complexity	2^4	2^8	2^{18}	2^{26}	2^{35}	2^{36}	2^{43}	2^{44}	2^{51}	2^{52}	2^{58}
number of rounds	4	6	8	9	10	11	12	13	14	15	16

tables. However, this does not mean that this is the only available description of the DES algorithm.

Several researchers have attempted to provide alternative descriptions of the algorithm for a number of reasons. The description of the DES algorithm based on tables is not very transparent; it does not provide much insight into the characteristics of the algorithm. Alternative descriptions may shed some light on how the algorithm actually functions. Also, a better understanding of the algorithm enables structured research into its resistance to crypt-analytic attacks. This could provide useful information on the strength of the algorithm. Furthermore, alternative descriptions could lead to simpler hardware implementations, which, in turn, could lead to cheaper and faster DES chips.

In this section we will present the results achieved by researchers at the Catholic University of Leuven and at the Philips Research Laboratory in Brussels (Davio *et al.* (1983)).

An iterative DES algorithm

Examining Figure 4.1 once again, we see that when we disregard the initial permutation IP and its inverse, the scheme is in fact an iterative process. Suppose we insert two new modules IP and IP^{-1} at each step. This would result in a cascade of 16 identical modules, with the small exception of the 16th step, where the left-hand and right-hand bits are not exchanged. Ignoring this dissymmetry for the time being, the system becomes purely iterative and can be realised by a sequential machine, as in Figure 4.5. The

Figure 4.5. An iterative scheme for DES.

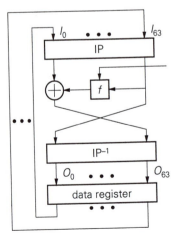

dissymmetry of the 16th step can easily be incorporated later by including switches which disable the interchanging module on the 16th step.

Models of the DES based on transformations

The above has demonstrated that the DES algorithm can be considered as an iterative algorithm. It is clear that this enables a simpler hardware implementation of the algorithm. In this section we will examine alternative descriptions of DES, based on transformations. First, consider Figure 4.6.

Figure 4.6(a) shows that two n-input n-output functions f after a branch have the same effect as a single function f, placed before the branch. This can easily be verified for the initial permutation IP (and its inverse), which is, in fact, a 64-input 64-output function. It is left to the reader to check that this equivalence also applies to the case in which the number of inputs is not equal to the number of outputs, such as the expansion operation E.

Figure 4.6(b) depicts a trivial equivalence; a function φ followed by its inverse function will result in the original input values at the outputs. This means that a cascade of two inverse modules may be inserted at any arbitrary position in the scheme. For instance, IP followed by IP^{-1}.

Finally, Figure 4.6(c) shows that when the binary outputs of two identical functions π are added together, the result is the same as when the inputs are added first, before the function. This statement holds as long as the function

Figure 4.6. Elementary transformations.

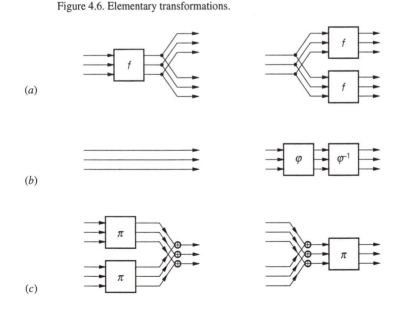

(a)

(b)

(c)

Figure 4.7. Transformation in which the permutation P becomes redundant.

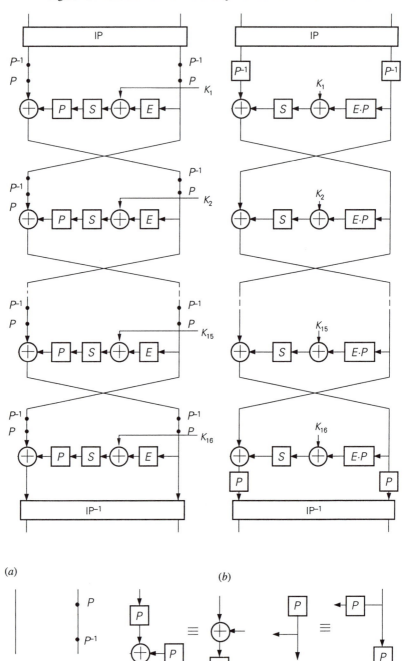

(a)   (b)

(c)

π is linear, i.e. provided:

$$\pi(a) + \pi(b) = \pi(a + b).$$

We can easily verify that the initial permutation and the expansion operation satisfy this condition.

Figure 4.7(a) is a more abstract representation of Figure 4.1, in which the

Figure 4.8. Transformations of the 48-bit model of the DES.

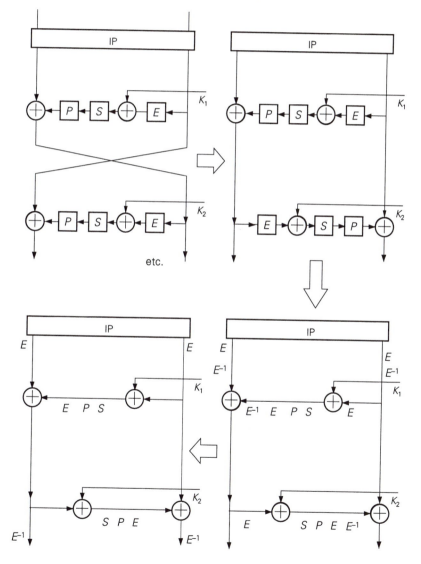

cipher function has been separated into the expansion operation E, the S-boxes and the permutation P. In addition, two modules P and P^{-1} have been inserted before each processing step. The transformations of Figure 4.6 tell us that the characteristics of the algorithm are not altered by this. Figure 4.7(c) shows the transformations which are used now.

Figure 4.9. The 48-bit model of the DES.

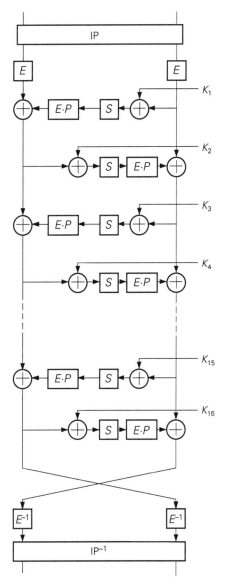

When these transformations are inserted at the right instance, the scheme of Figure 4.7(*b*) is equivalent to that of Figure 4.7(*a*). In Figure 4.7(*b*), the permutations P and P^{-1} are positioned at the beginning and the end of the scheme in such a way that makes it possible to combine these with the initial permutation IP to give a single function. This is also true for E and P. We can therefore conclude that the table P has become redundant, which makes the DES algorithm less complex than our first impression.

Of course, other transformations may also be introduced. The following description shows that the DES algorithm can also be examined with a 48-bit model, unlike the previous descriptions, which were all based on blocks of 2×32 bits. Figure 4.8 shows part of the general scheme. Suitable transformations based on those of Figure 4.6 can be used to simplify this scheme, eventually resulting in that of Figure 4.9. The output of the expansion operation is a 48-bit word, so therefore this scheme will process 48 bits instead of the original 2×32 bits. Finally, we can conclude that this implementation requires one table less than originally specified by the designers.

4.4 Analysis of the DES

In addition to alternative descriptions of the DES, researchers are also interested in the possibility of describing the DES algorithm by analytic functions. Once a cipher algorithm can be described entirely by a number of functions, it has, in fact, been 'cracked'. With a known-plaintext-attack, the cryptanalyst can compute a system of functions, which should lead to the unknown key.

So far, it has not proved possible to provide a comprehensive analytic description of the DES and it seems as if this is unrealisable. Analytic functions have been found only for parts of the algorithm. We will examine these here as this will provide further insight into the operation of the algorithm.

Analysis of the initial permutation

Consider the modified matrix IP; this is the matrix as given in Section 4.1, in which the numbers run from 0 to 63, rather than from 1 to 64:

57	49	41	33	25	17	9	1
59	51	43	35	27	19	11	3
61	53	45	37	29	21	13	5
63	55	47	39	31	23	15	7
56	48	40	32	24	16	8	0
58	50	42	34	26	18	10	2
60	52	44	36	28	20	12	4
62	54	46	38	30	22	14	6

Modified matrix IP:

This modified matrix IP transforms 64 input bits $\{X_0, X_1, \ldots, X_{63}\}$ to 64 output bits $\{Y_0, Y_1, \ldots, Y_{63}\}$. With the given matrix we find that $Y_0 = X_{57}$, $Y_1 = X_{49}$ etc.

Examination of the relations between the indices of the input and output bits reveals the following. If these indices are represented by a binary notation, we find for the separate bits of the indices:

$$(y_5, y_4, y_3, y_2, y_1, y_0) = (\underline{x_0}, x_2, x_1 \underline{x_5}, \underline{x_4}, \underline{x_3}), \qquad (4.9)$$

in which the underlined bits are the inverted values of the original bits. This is illustrated by the following example.

Example

Consider input bit X_2. The index of this bit is represented in binary form by (000010). The above formula states that for the bits of the index of the corresponding output bit, y_0 is the inverse of bit x_3, y_1 the inverse of x_4 etc. So, if the value (000010) is processed according to eq. (4.9), we find:

$$(y_5, y_4, y_3, y_2, y_1, y_0) = (\underline{00}1\underline{000}) = (101111) = 47.$$

Apparently, $Y_{47} = X_2$, which agrees entirely with the relation which is given by the modified matrix IP. $\triangle$

By representing the remaining indices of the input bits in binary form, the indices of the output bits can be computed in a similar manner.

Returning to the modified matrix IP, we can draw another conclusion. The fourth and fifth rows of the matrix show that when the input consists of ASCII characters, all parity bits will end up in one output byte. Also, the 32 left-hand bits consist of all the even bits and the 32 right-hand bits consist of all the odd bits.

Analysis of the expansion operation E

The expansion operation E converts 32 input bits into 48 output bits. Denoting the input bits by $(r_0, r_1, \ldots, r_{31})$ and the output bits by $(v_0, v_1, \ldots, v_{47})$, the expansion operation will result in:

$$v_{6i+j} = r_{4i-1+j} \pmod{32}, \tag{4.10}$$

in which

$$i \in (0,\dots,7),$$
$$j \in (0,\dots,5).$$

Example

Considering v_{23}, so $6i + j = 23$, we get $i = 3$ and $j = 5$, from which it follows that $4i - 1 + j = 16$, so eventually $v_{23} = r_{16}$. Figure 4.3 confirms that this is correct. $\triangle$

Now that we have found analytic expressions for both the initial permutation and the expansion operation, we can investigate these operations together. Examine Figure 4.10, which should be read from bottom to top.

Assume the input of the first S-box is given by $(I_0,I_1,\dots,I_5)$. These six bits are the result of the binary addition of the first six bits of a subkey and the first six bits of the output of the expansion operation. The first 6 bits of the output of the expansion operation are given by $(v_0,v_1,\dots,v_5)$. According to

Figure 4.10. Analysis of IP and E together.

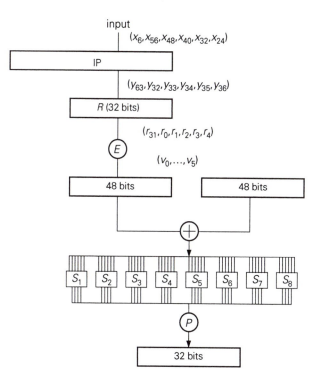

eq. (4.10), the corresponding input of the expansion operation must have been equal to $(r_{31},r_0,r_1,r_2,r_3,r_4)$. Since the first six input bits of the expansion operation are identical to the first six bits of the 32 right-hand bits after the initial permutation, $(r_{31},r_0,r_1,r_2,r_3,r_4)$ must be equal to $(Y_{63},Y_{32},Y_{33}, Y_{34},Y_{35},Y_{36})$. Based on eq. (4.9), we can then draw the conclusion that the sequence $(Y_{63},Y_{32},Y_{33},Y_{34},Y_{35},Y_{36})$ must originate from the sequence $(X_6,X_{56},X_{48},X_{40},X_{32},X_{24})$.

This analysis reveals that five of the six bits which form the input of the S-boxes must originate from the same byte positions of the original input. In this example this is the first position, as can be concluded from the indices 56, 48, 40, 32 and 24.

It can easily be demonstrated that this also holds for the inputs of the other S-boxes.

Analysis of S-boxes

In the previous section we managed to find expressions for the calculations performed by the DES algorithm, as far as the inputs of the S-boxes. However, this has not yet proved possible for the S-boxes themselves. No analytic expression has been found for the $8 \times 4 = 32$ functions which describe the S-boxes. Even so, we must still not exclude the possibility of a 'trapdoor', however unlikely this may be, with which the output of the S-box could be described in terms of the input. Investigations into this possibility have not yet produced any clear results, although the following observations can be made:

- The output of the S-box is a non-linear function of the input.
- Each input bit influences at least two output bits.
- The difference between the number of zeros and ones of the output is minimal when one input bit is kept constant.
- S-box 4 is redundant. This box can be described by a single function, instead of four different functions, as is illustrated in Figure 4.11.

Assuming $F1$ is the function which determines the 1st output bit, the remaining three output bits can also be calculated with $F1$ by adding the 6th input bit. In this manner the number of functions which describe the S-box is reduced from 32 to 29.

4.5 The modes of DES

Electronic code-book mode (ECB mode)

If we wish to implement the DES algorithm in practice, then the most obvious way is given by Figure 4.12. On the transmitter's side, blocks of 64 bits are enciphered and sent to the receiver, where they are deciphered and the original blocks are obtained.

If during transmission from one side to the other bit errors occur in a block, the block of data resulting after decipherment will be incorrect. Subsequent blocks of data, however, will not be affected. Luckily therefore, errors will not propagate.

The same applies to the case in which an entire block of data is lost. Again, this will not affect subsequent blocks. If the demarcation between blocks is obscured the data will be incorrectly deciphered. As soon as this is restored (synchronised), decipherment will be correct again.

A major disadvantage of the DES in the ECB mode is that the security of the system relies entirely on the secrecy of the key. Once cryptanalysts have managed to obtain the key, they can decrypt the ciphertext into plaintext. This mode therefore imposes strict requirements with respect to guarding the secrecy of the key.

Further attention must be paid to the fact that in this mode there is always a one-to-one relation between the input and the output; a given input will

Figure 4.11. Redundancy in *S*-box 4.

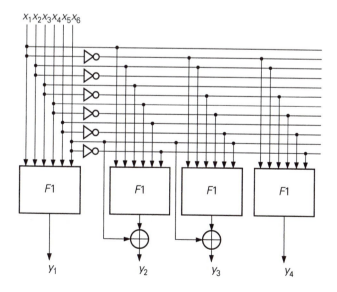

always produce the same output. This is why this mode is called the 'electronic code-book' mode; it is theoretically possible to construct a code-book which describes the one-to-one relation between input and output.

For the above reasons the DES algorithm is mostly used in other modes, which allow for a better security. These modes are:

– the cipher block chaining (CBC) mode;
– the k-bit cipher feedback (CFB) mode;
– the k-bit output feedback (OFB) mode.

Cipher block chaining mode (CBC mode)

The CBC mode of DES is depicted in Figure 4.13. Before the actual DES encipherment takes place, the previous block of ciphered data is added to the plaintext. For the first block of data a so-called *initial vector* (IV), a block of 64 random bits is used. When the blocks are deciphered, they are again added in binary form to the initial vector and the previous cipher block, which results in the original block of plaintext.

The initial vector can be regarded as an extra key which guarantees extra security. If the enciphered blocks of data are denoted by $C(i)$ and the block of plaintext by $M(i)$, then:

Figure 4.12. Electronic code-book mode of DES.

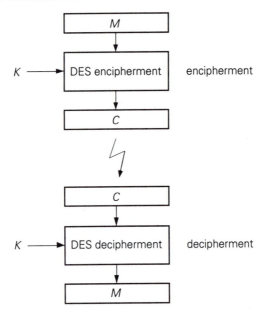

encipherment:

$$C(1) = \text{DES }[M(1) + \text{IV}],$$
$$C(2) = \text{DES }[M(2) + C(1)],$$
$$C(i) = \text{DES }[M(i) + C(i\text{-}1)], \qquad i > 2;$$

decipherment:

$$M(1) = \text{DES }[C(1)] + \text{IV},$$
$$M(2) = \text{DES }[C(2)] + C(1),$$
$$M(i) = \text{DES }[C(i)] + C(i\text{-}1), \qquad i > 2,$$

where DES [.] indicates that the block between the brackets is processed by the DES algorithm.

We can easily examine the effect of bit errors and synchronisation errors with the given formulae.

Bit errors

Suppose the receiver receives $C(1) + Q$, in which Q is a given bit error sequence, instead of $C(1)$. This will result in the following expressions for the deciphered blocks of plaintext:

$$M(1) = \text{DES }[C(1) + Q] + \text{IV},$$
$$M(2) = \text{DES }[C(2)] + C(1) + Q,$$

Figure 4.13. CBC mode of DES.

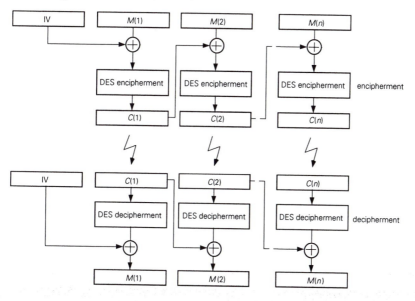

$$M(i) = \text{DES } [C(i)] + C(i\text{-}1), \qquad i > 2.$$

Clearly, the first two blocks will be incorrect, but subsequent blocks will be deciphered correctly.

Missing blocks
Suppose block $C(1)$ was not received. Then it follows that:

$$M(1) = \text{DES } [C(2)] + \text{IV},$$
$$M(2) = \text{DES } [C(3)] + C(2),$$
$$M(i) = \text{DES } [C(i\text{+}1)] + C(i), \qquad i > 2.$$

After one block autosynchronisation occurs. $M(1)$ is incorrect, but the following blocks are deciphered correctly.

Lost block boundaries
The above formulae also reveal that if the received demarcation between blocks is in the wrong position, the decipherment will be incorrect until the correct boundaries between blocks have once again been restored.

k-bit cipher feedback mode (CFB mode)

In the CFB mode the ciphertext is generated by adding k bits modulo 2 to the first k bits of the output of the DES algorithm. Referring to Figure 4.14, we see that the ciphertext of k bits is used as the input for the DES algorithm. Furthermore, we see that this mode also requires an initial vector

Figure 4.14. k-bit CFB mode of DES.

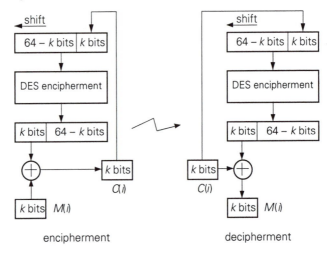

encipherment   decipherment

IV. Assuming this vector also consists of 64 bits (although this is not strictly necessary), then the input for the first round of the DES algorithm will be $(64 - k)$ zeros followed by the k bits of IV. This is denoted by [0,IV]. For the second round, the k bits of the ciphertext $C(1)$ are placed at the last k positions of the input block of DES and the remaining bits of the input are shifted k positions to the left. Thus, the input becomes a sequence of $(64 - 2k)$ zeros followed by the k bits of the IV and k bits of $C(1)$: [0,IV,$C(1)$] etc. Obviously, after a while first the zeros, then IV, $C(1)$ etc. will disappear from the input of DES as they are repeatedly shifted to the left.

Encipherment and decipherment is described by the following expressions:

encipherment:
$$C(1) = M(1) + k\mathrm{DES}\ [0,\mathrm{IV}],$$
$$C(2) = M(2) + k\mathrm{DES}\ [0,\mathrm{IV},C(1)],$$
$$C(3) = M(3) + k\mathrm{DES}\ [0,\mathrm{IV},C(1),C(2)],$$
etc.;

decipherment:
$$M(1) = C(1) + k\mathrm{DES}\ [0,\mathrm{IV}],$$
$$M(2) = C(2) + k\mathrm{DES}\ [0,\mathrm{IV},C(1)],$$
$$M(3) = C(3) + k\mathrm{DES}\ [0,\mathrm{IV},C(1),C(2)],$$
etc.

in which $k\mathrm{DES}$ means that only the first k bits of the output of DES are regarded.

The influence of bit errors and synchronisation errors will now be analysed.

Bit errors
If we receive $C(1) + Q$, instead of $C(1)$, this will lead to the following expressions for the blocks of plaintext:

$$M(1) = C(1) + Q + k\mathrm{DES}\ [0,\mathrm{IV}],$$
$$M(2) = C(2) + k\mathrm{DES}\ [0,\mathrm{IV},C(1) + Q],$$
$$M(3) = C(3) + k\mathrm{DES}\ [0,\mathrm{IV},C(1) + Q,C(2)],$$
etc.

Clearly, errors will propagate until the k bits of $C(1) + Q$ are no longer a part of the input block of the DES algorithm.

Missing blocks
If the receiver fails to receive block $C(1)$, then:

$$M(1) = C(2) + kDES\ [0,IV],$$
$$M(2) = C(3) + kDES\ [0,IV,C(2)],$$
$$M(3) = C(4) + kDES\ [0,IV,C(2),C(3)],$$
etc.

In other words, the ciphertexts will be deciphered correctly, as soon as the input blocks of DES no longer contain IV.

Lost block boundaries
If the position of the delimiters between blocks is lost, then the input of the DES algorithm will be incorrect. Once the system is synchronised and the input of the DES is free of these incorrect bits, correct decipherment will be restored.

k-bit output feedback mode (OFB mode)

In the OFB mode the DES algorithm is not applied directly to the plaintext itself. Instead, it is used to produce bit sequences which are added to the plaintext. This is depicted in Figure 4.15. Again an initial vector IV is used. After the IV is encrypted, k bits of the output are fed back to the input of the DES algorithm and so forth.

Encryption and decipherment can be described as follows:

Figure 4.15. k-bit OFB mode of DES.

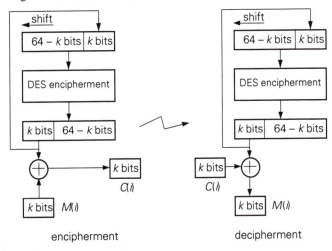

encipherment:

$$C(1) = M(1) + k\text{DES } [0,\text{IV}] = M(1) + T(1),$$
$$C(2) = M(2) + k\text{DES } [0,\text{IV},T(1)] = M(2) + T(2),$$
etc.;

decipherment:

$$M(1) = C(1) + T(1),$$
$$M(2) = C(2) + T(2),$$
etc.,

in which by definition, with $i > 1$, $T(i)$ is equal to:

$$T(i) = k\text{DES } [...,T(i-1)].$$

These formulae lead to the conclusion that no error propagation will occur and that when the block demarcations are lost, correct decipherment is restored as soon as the boundary of a block has been found again. It is left to the reader to verify that when the beginning of a block is lost, the data will be incorrectly deciphered.

In practice, which mode of DES is most suitable depends on the kind of application. For each individual application we must investigate whether a certain mode satisfies the requirements of the application and its environment. For instance, if the DES is used for the cryptographic protection of a satellite communications link, which is easily aflicted by bit errors, the OFB mode of the DES is preferred, as this mode guarantees correct decipherment of the other bits, etc.

In Chapter 7 we will turn our attention to the specific properties of the DES modes with regard to authentication.

4.6 Future of the DES

So far, the analysis of the DES has revealed that only certain functions of the algorithm can be described by analytic expressions. Several properties of the S-boxes have been discovered, including the redundancy of one S-box, but a 'trapdoor' (assuming one exists) has not been found.

Ultimately, the only option which lies open to a potential hacker is an extensive key search. However, the financial investment which enable the key to be found within a reasonable period of time is considerable.

In practice, the security of the system is sometimes increased further by implementing multiple DES algorithms. This should only be done when the necessary care is taken.

In Figure 4.16 the plaintext is enciphered twice with two independent keys $K1$ and $K2$. Contrary to what may be expected, this does not lead to an effective key length of 112 bits. Suppose that after a known-plaintext-attack we have two pairs of plaintexts and ciphertexts: $(M1,C1)$ and $(M2,C2)$. $M1$ is enciphered with all possible keys and $C1$ is deciphered with all possible keys. This will result in two sets of data: a set $E_{Ki}(M-1)$ and a set $D_{Kj}(C1)$, in which Ki and Kj represent all possible keys. The elements of these sets are then compared with each other until $E_{Ki}(M1)= D_{Kj}(C1)$. We now have two keys for which:

$$C1 = E_{Kj}(E_{Ki}(M1)).$$

Theoretically, there are 2^{48} pairs of keys including the pair $(K1,K2)$ for which we are looking. By generating a similar set for $(M2,C2)$ of pairs of potential keys, the search for $(K1,K2)$ can be restricted even further. This kind of attack is called a *meet-in-the-middle-attack*. The complexity of this approach is comparable to an exhaustive key search in a key space of 2^{58} elements (Merkle and Hellman, 1981). This value is only marginally larger than that for a single encipherment and nowhere near the desired 2^{112}.

A much better result is obtained by enciphering the plaintext with $K1$, deciphering it with $K2$ and then enciphering it with $K1$ again (see Figure 4.17(a)). Note that when $K1 = K2$ this is the same as a single encipherment. This threefold encipherment can also be subjected to a meet-in-the-middle

Figure 4.16. Twofold encipherment based on the DES.

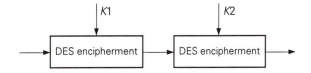

Figure 4.17(a). Threefold encipherment with the DES; (b) triple-DES.

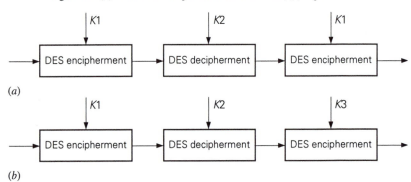

(a)

(b)

attack. However, unlike the previous method, where a couple of pairs of plaintext and ciphertext were sufficient, we must now have a very large number of plaintexts. Therefore, this kind of attack no longer represents a serious threat to this approach. The complexity is nowadays of the order of 2^{80}.

The best result is obtained when the method of Figure 4.17(b) is used; encipherment with $K1$, decipherment with $K2$ and finally encipherment with $K3$. This procedure is often referred to as triple-DES.

4.7. IDEA (International Data Encryption Algorithm)

A number of other cryptographic algorithms have been developed to offer an alternative to the DES. One of these algorithms is called IDEA (International Data Encryption Algorithm), which was developed by Lai (1991), originally under the name IPES. IDEA is currently used within PGP (Pretty Good Privacy), the cryptographic system for Internet and E-mail security.

IDEA also works with blocks of 64 bits, jut as DES does, although now, each block is divided internally into 4 blocks of 16 bits each. The number of rounds is 8 and the size of the key is 128 bits.

The algorithm is illustrated in Figure 4.18. The input of the IDEA algorithm consists of 4 blocks of 16 bits each, denoted by $X1$, $X2$, $X3$ and $X4$. In every round, 6 subkeys are used, each 16 bits long. For round i, these are denoted by $K_{i,1},...,K_{i,6}$. Since there are 8 rounds, 48 subkeys are used, plus 4 extra keys, which are used after the last round to transform the output. The 4 output blocks are denoted by $Y1$, $Y2$, $Y3$ and $Y4$.

In each round, the 16 bits blocks are XOR-ed, added and multiplied as indicated in the figure. The multiplication modulo $2^{16}+1$ can be regarded as the substitution box of IDEA. After the last step, each of the resulting 16-bit blocks are multiplied modulo $2^{16}+1$ by its corresponding subkey.

The 52 subkeys are generated in a very straightforward manner. The original key of 128 bits is divided into 8 blocks of 16 bits. These blocks represent the subkeys $K_{1,1},...,K_{1,6}$, which are used in the first round and two subkeys $K_{2,1}$ and $K_{2,2}$, which represent the first two keys of the second round. The bits are then shifted 25 positions to the left and again, they are divided into blocks of 16 bits, resulting in 8 new subkeys: four for the second round and four for the third round. Again, they are shifted another 25 bits to the left, and so on.

The same algorithm can be used for deciphering, just as with DES. The keys must also be supplied in reverse order for IDEA, although they must be first modified. The subkeys used for deciphering are the inverses of the encipherment keys for multiplications as well as additions. The first of these operations is performed modulo $2^{16}+1$; the second is performed modulo 2^{16}.

Figure 4.18. IDEA.

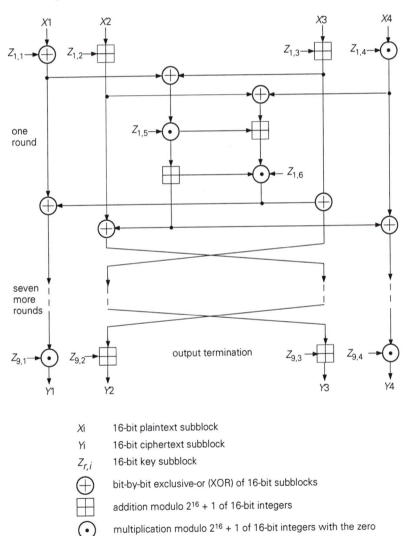

Xi	16-bit plaintext subblock
Yi	16-bit ciphertext subblock
$Z_{r,i}$	16-bit key subblock
⊕	bit-by-bit exclusive-or (XOR) of 16-bit subblocks
⊞	addition modulo $2^{16} + 1$ of 16-bit integers
⊙	multiplication modulo $2^{16} + 1$ of 16-bit integers with the zero subblock corresponding to 2^{16}

Compare Table 4.7.

As far as the use of the IDEA is concerned, it can be used in the same modes as the DES. From a practical point of view the IDEA is an attractive algorithm. It is approximately twice as fast as the DES algorithm.

Presently, no successful cryptanalytic attacks on the IDEA have been recorded. Although the number of rounds for the IDEA is only half that of the DES, the IDEA algorithm can still safely withstand an attack based on differential cryptanalysis.

The only method of attack that remains is to try all possible keys (*exhaustive key search*). Since the key consists of 128 bits, the complexity of finding the key is $2^{128} \approx 10^{38}$. In order to find the solution within a day, we would need, for example, at least 10^{25} ICs running in parallel, each capable of trying 100 million keys a second. However, there is not enough silicon on earth to make that number of ICs and we are limited by practical and physical restrictions.

Table 4.7. Subkeys for encipherment and decipherment.

round	encryption subkeys	decryption subkeys
1	$Z_{1,1}\,Z_{1,2}\,Z_{1,3}\,Z_{1,4}\,Z_{1,5}\,Z_{1,6}$	$Z_{9,1}^{-1}\,{-}Z_{9,2}\,{-}Z_{9,3}\,Z_{9,4}^{-1}\,Z_{8,5}\,Z_{8,6}$
3	$Z_{2,1}\,Z_{2,2}\,Z_{2,3}\,Z_{2,4}\,Z_{2,5}\,Z_{2,6}$	$Z_{8,1}^{-1}\,{-}Z_{8,3}\,{-}Z_{8,2}\,Z_{8,4}^{-1}\,Z_{7,5}\,Z_{7,6}$
3	$Z_{3,1}\,Z_{3,2}\,Z_{3,3}\,Z_{3,4}\,Z_{3,5}\,Z_{3,6}$	$Z_{7,1}^{-1}\,{-}Z_{7,3}\,{-}Z_{7,2}\,Z_{7,4}^{-1}\,Z_{6,5}\,Z_{6,6}$
4	$Z_{4,1}\,Z_{4,2}\,Z_{4,3}\,Z_{4,4}\,Z_{4,5}\,Z_{4,6}$	$Z_{6,1}^{-1}\,{-}Z_{6,3}\,{-}Z_{6,2}\,Z_{6,4}^{-1}\,Z_{5,5}\,Z_{5,6}$
5	$Z_{5,1}\,Z_{5,2}\,Z_{5,3}\,Z_{5,4}\,Z_{5,5}\,Z_{5,6}$	$Z_{5,1}^{-1}\,{-}Z_{5,3}\,{-}Z_{5,2}\,Z_{5,4}^{-1}\,Z_{4,5}\,Z_{4,6}$
6	$Z_{6,1}\,Z_{6,2}\,Z_{6,3}\,Z_{6,4}\,Z_{6,5}\,Z_{6,6}$	$Z_{4,1}^{-1}\,{-}Z_{4,3}\,{-}Z_{4,2}\,Z_{4,4}^{-1}\,Z_{3,5}\,Z_{3,6}$
7	$Z_{7,1}\,Z_{7,2}\,Z_{7,3}\,Z_{7,4}\,Z_{7,5}\,Z_{7,6}$	$Z_{3,1}^{-1}\,{-}Z_{3,3}\,{-}Z_{3,2}\,Z_{3,4}^{-1}\,Z_{2,5}\,Z_{2,6}$
8	$Z_{8,1}\,Z_{8,2}\,Z_{8,3}\,Z_{8,4}\,Z_{8,5}\,Z_{8,6}$	$Z_{2,1}^{-1}\,{-}Z_{2,3}\,{-}Z_{2,2}\,Z_{2,4}^{-1}\,Z_{1,5}\,Z_{1,6}$
output	$Z_{9,1}\,Z_{9,2}\,Z_{9,3}\,Z_{9,4}$	$Z_{1,1}^{-1}\,{-}Z_{1,2}\,{-}Z_{1,3}\,Z_{1,4}^{-1}$

5

Shift registers

5.1 Stream and block enciphering

In the previous chapter we saw that the DES algorithm enciphers blocks of 64 bits. This algorithm is an example of what is called *block enciphering*. For block encipherment, a fixed number of elements of the plaintext are enciphered together as a whole. A transposition cipher can therefore also be considered as an example of block enciphering.

In addition to block enciphering, we also find *stream enciphering*. Here, the plaintext is regarded as a collection of separate elements. These elements can represent the letters of a text and also the zeros and ones of bit stream. Stream enciphering is characterised by the fact that it encrypts the plaintext element by element.

One way of achieving this is the one-time path (see Figure 5.1(a)), which we encountered in Section 3.2. A random, i.e. unpredictable, sequence K is generated, which in the binary case will simply consist of zeros and ones. The outcome of repeatedly tossing a coin, for instance, would suffice. This sequence is added to the binary representation of the plaintext M, which will result in the ciphertext C. Thus:

$$C = M + K \ (\text{mod } 2), \tag{5.1}$$

assuming the length of K is at least as great as the length of the plaintext. Since K is a random sequence, C will also have a random character. This implies that for a given ciphertext each decipherment has the same probability.

The receiver deciphers the text in a manner similar to the method of encipherment:

$$M = C + K \pmod 2 \qquad\qquad (5.2)$$

Naturally, the sequence K must also be available to the receiver.

In practice, it may be inconvenient to generate or transport an entire bit stream K, in which case shift registers can offer a useful solution. These are capable of generating so-called *pseudorandom* sequences, i.e. series of zeros and ones which have a nearly unpredictable or random structure. Figure 5.1(*b*) illustrates the use of shift registers in an implementation of a stream encipherment.

A shift register, or in more general terms, a pseudorandom generator will produce a pseudorandom sequence K which depends on the initial value loaded into the shift register and a given key K' of restricted length. This sequence K is added to the binary plaintext M. Both the transmitter and the receiver must have the same initial parameters in order to guarantee that the same sequences can be generated on both sides.

Shift registers have the advantage that they are relatively cheap. Also, they can generate long bit streams from a small number of initial parameters. They are therefore extremely suitable for practical applications. In addition, they can easily operate at speeds above 20 Mbit/s, thus enabling real-time encipherment of speech etc. For these reasons, pseudorandom series generated by shift registers are often used for protecting data during transportation.

Another advantage of stream encipherment is the fact that it is relatively insensitive to errors introduced during transportation. If a bit of the ciphertext is modified during transportation then only this bit will be deciphered incorrectly when the text reaches the receiver. However, a single

Figure 5.1(*a*). One-time path, (*b*) enciphering and deciphering with the aid of pseudorandom sequences.

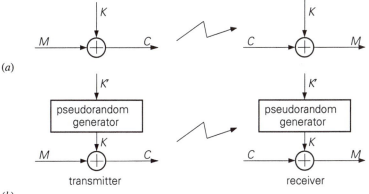

(*a*)

(*b*)

bit error in data which is encrypted with a block encipherment will cause the entire block to be deciphered incorrectly, since each element of the output block depends on all the elements of the corresponding input block. In some situations this may prove to be an advantage; for instance, if intruders deliberately modify a text, this will be easier to detect if the text is encrypted with a block cipher.

5.2 The theory of finite state machines

In this and the following sections we will examine various methods of generating pseudorandom sequences. First, however, we must pay attention to the theory of '*finite state machines*'. Basically, these machines are capable of producing sequences of symbols from other sequences.

We will base our description of finite state machines on the following sets and functions:

a set of states	$K = \{k_i\}$,
a set of input symbols	$A = \{a_i\}$,
a set of output symbols	$B = \{b_i\}$,
state function	$\delta\colon K_{n+1} = \delta(K_n, a_n)$,
output function	$\mu\colon b_n = \mu(K_n, a_n)$.

The state function expresses the fact that every new state depends on both the previous state and the given input symbol. The output function states that each output is determined by the current state of the machine and the input symbol. With each input symbol of a sequence of input symbols, the machine will attain a unique state and thus generate a different output symbol each time.

This is illustrated by the following. Assume the input of our finite state machine is a series of zeros and ones and that the output symbols are given by the set $B = (\alpha, \beta, \gamma)$. The machine has three states, R, S and T, respectively. Finally, the state functions and output functions which lead to the relations between the states, the input symbols and the output symbols can be summarised by the following function table:

	0	1
R	S, β	T, α
S	R, γ	S, γ
T	T, γ	R, β

This table is interpreted as follows. When the finite state machine is in state *R* and the input symbol is 0, the machine will pass into state *S* and generate the output symbol *β*. However, if the input symbol had been a 1, the machine would have passed into state *T*, and generated the output symbol *α*.

The following table demonstrates what happens when the sequence (00101100) is supplied to the input of the machine.

time	input sequence	state	output sequence
1	0	R	β
2	0	S	γ
3	1	R	α
4	0	T	γ
5	1	T	β
6	1	R	α
7	0	T	γ
8	0	T	γ
9		T	

In addition to the function table, another means of describing a finite state machine is provided by a so-called *state diagram*. The state diagram of the above example is given in Figure 5.2. Each node represents a state. The arrows indicate a transition from one state to another with a given input symbol. The symbol in brackets appears at the output of the machine as it passes from one state to the other.

An alternative state diagram for the above example is given in Figure 5.3. The diagram has now been modified by adding an extra state *U* to the diagram, in addition *R*, *S* and *T*. However, it can easily be verified that the same sequence of input symbols will still generate the same sequence of

Figure 5.2. Example of a state diagram.

Figure 5.3. State diagram of Figure 5.2, with 1 additional state.

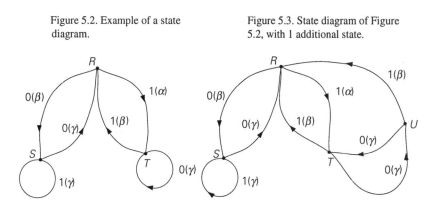

output symbols. In other words, the state diagrams in Figures 5.2 and 5.3 define the same input–output relations. In general, we will attempt to define the behaviour of the finite state machine with the most straightforward representation and a minimal number of states. Therefore, the state diagram of Figure 5.2 is preferred to that of Figure 5.3.

In order to test the characteristics of a finite state machine, a number of characteristic sequences of input symbols can be supplied to the machine. Several examples of such sequences are given below:

(1) all zeros: 00 0
(2) all ones: 11 1
(3) a one followed by all zeros: 10 0
(4) several ones followed by zeros: 11 ...   100   ... 0

Speaking in electrical terms, we can say that the first two series can be considered as a DC input voltage and that sequences 3 and 4 correspond to an impulse and a block wave, respectively.

If the input remains constant, the output will eventually become periodic. In other words, the sequence of output symbols will be repeated after some time.

Assuming the initial state of the machine in our example is state R, the above sequences of input symbols will result in the following output:

(1) input sequence A,   output sequence $\beta\gamma$ $\beta\gamma$ $\beta\gamma$ $\beta\gamma$ $\beta\gamma$...
                                 states RS RS RS RS RS...

(2) input sequence B,   output sequence $\alpha\beta$ $\alpha\beta$ $\alpha\beta$ $\alpha\beta$ $\alpha\beta$...
                                 states RT RT RT RT RT...

(3) input sequence C,   output sequence $\alpha\gamma$ $\gamma\gamma$ $\gamma\gamma$ $\gamma\gamma$ $\gamma\gamma$...
                                 states RT TT TT TT TT

(4) input sequence D
   (uneven number of ones)   output sequence $\alpha\beta$ $\alpha\beta$ α .$\gamma\gamma$...
                                 states RT RT R.. TT...

   (even number of ones)   output sequence $\alpha\beta\alpha\beta$...$\beta\gamma\beta$...
                                 states $RTRT$...RSR...

5.3. Shift registers

Figure 5.4 shows the general structure of a so-called feedback shift register (FSR), consisting of m sections. Each section can hold one bit. The shift register is controlled by a clock input. On every pulse of the clock the bits are shifted one section to the right. The bits which are generated at section 0 form the output of the shift register. Section 5.1 demonstrated that such a sequence can be used to encipher the plaintext by simply adding the two in binary.

As the bits are shifted to the right, it is necessary to supply new bits to section $m - 1$. These bits can be obtained from a feedback loop, containing a module which calculates the values of the new bits, according to a so-called *feedback function*, based on the values which are fed back from the outputs of the other sections.

In the previous section we examined finite state machines. From Figure 5.4, we can conclude that the feedback shift register can be considered as a special type of finite state machine. A state of the register is defined by a given sequence representing the contents of the separate sections. Every clock pulse causes a transition to a new state, which is determined by the previous state and the feedback function. It is clear that if there are m sections, there will be 2^m possible states. And since there are 2^m states, there will be 2^{2^m} possibilities for the feedback function. This can be explained as follows. The feedback function can be any possible non-linear function of $x_0, x_1, \ldots, x_{m-1}$, where x_0 represents the contents of section 0, x_1 the contents of section 1, etc. The terms of the function are selected from the set which is given by:

$$(x_0, x_1, \ldots, x_{m-1}, x_0 x_1, \ldots, x_{m-1} x_{m-1}, \ldots, x_0 x_1 \ldots x_{m-1}).$$

There are $\binom{m}{1}$ terms consisting of one element, $\binom{m}{2}$ terms comprising two elements, etc. and finally one term consisting of m elements (the product $x_0 x_1 \ldots x_{m-1}$). In addition, one term can be equal to 1, to allow for the inverse value. Thus, the total number of possible terms of which the feedback function may consist, is:

Figure 5.4. Schematic of a feedback shift register.

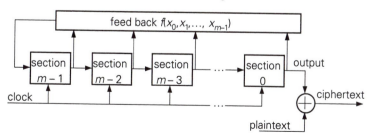

$$\binom{m}{0} + \binom{m}{1} + \binom{m}{2} + \ldots + \binom{m}{0} = 2^m.$$

Each feedback function is given by specifying which of the 2^m terms appear in the function. Since for each term there are two possibilities (it either does or does not appear in the function), this finally results in 2^{2^m} possible feedback functions.

A *linear-feedback shift register* (LFSR) is defined as a feedback shift register whose feedback function $f(x_0, x_1, \ldots, x_{m-1})$ is a linear function, i.e., the feedback function can be expressed as:

$$f(x_0, \ldots, x_{m-1}) = c_0 x_0 + c_1 x_1 + \ldots + c_{m-1} x_{m-1}$$

$$= \sum_{i=0}^{m-1} c_i x_i. \tag{5.3}$$

The coefficients $c_0 - c_{m-1}$ can assume two values, 0 or 1, and thus determine whether or not a certain section is connected to the feedback loop. Therefore, there are 2^m linear functions in total.

When linear shift registers are used for enciphering plaintexts, the key consists of the initial vector, i.e. the initial state of the shift register plus the values of the coefficients $c_0 - c_{m-1}$. The receiver must therefore have access to this information in order to generate the pseudorandom sequence necessary for deciphering the ciphertext.

Figure 5.5 provides an example of a LFSR comprising three sections. The initial state is 111. The feedback function is defined as the binary addition of the contents of sections 0 and 1. The figure also shows the output of this shift register for the given initial state. Note that after six clock pulses, the shift register will return to its initial state and that after the first seven symbols the output will be repeated. Thus, we can speak of a period of 7. In general, the maximum period of a LFSR consisting of m sections, is $2^m - 1$. This is not quite equal to the number of possible states, i.e. 2^m, because the state 000 is an exception. If the LFSR passes into state 000, each subsequent state will also be 000, regardless of the feedback function and the period becomes 1. Hence, the maximum period is $2^m - 1$ and for the given example, the period is $2^3 - 1 = 7$.

Unfortunately, LFSRs are not always useful, since they are not resistant to a given-plaintext-attack, as we will see later. As an alternative, non-linear-feedback shift registers with a non-linear feedback function are often used. An example of a non-linear-feedback shift register is depicted in Figure 5.6. Here, the feedback function is given by:

$$f(x_2, x_1, x_0) = x_0 x_1.$$

The figure also explains how a subsequent state can be determined for this feedback function and a given state. The state diagram consists of two halves. If the initial state is (111), then the shift register will continuously produce ones at the output. However, all other initial states will eventually lead to a state in which only zeros appear at the output (note that from this point onwards ciphertext = plaintext).

Since multiplication is an asymmetrical operation, in applications based on non-linear shift registers we must ensure that the number of zeros in the plaintext is not too large compared to the number of ones.

According to Figure 5.6, the state (010) has two possible preceding states, i.e. (101) and (100). This is characteristic of non-linear shift registers. It can be demonstrated that for linear shift registers, each state is preceded by one state only; the state diagram will have no branches in this case.

In accordance with Figure 5.6, a non-linear shift register with a length of 3 can assume one of $2^3 = 8$ states and, in general, we find that a non-linear shift register of m sections will have 2^m possible states.

Figure 5.7 shows the so-called *de Bruyn graphs* for the cases $m = 1, ..., 5$, named after the Dutch mathematician N. G. de Bruyn. Each diagram of given order m shows all possible states and all possible transitions from one state to another. An output sequence can be determined by following a path

Figure 5.5. LFSR with three sections.

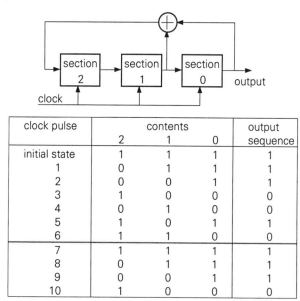

clock pulse	contents			output
	2	1	0	sequence
initial state	1	1	1	1
1	0	1	1	1
2	0	0	1	1
3	1	0	0	0
4	0	1	0	0
5	1	0	1	1
6	1	1	0	0
7	1	1	1	1
8	0	1	1	1
9	0	0	1	1
10	1	0	0	0

through the diagram and considering the superposition of successive states. Each state has two possible following states and the total number of possible states is 2^m. Thus, there are 2^{2^m} different sequences with a maximum period of 2^m for a non-linear-feedback shift register.

These diagrams show remarkable similarities to those which are used for describing Markov chains. It can easily be verified that a de Bruyn graph of order m is, in fact, identical to the state diagram of a $(m - 1)$ order Markov process.

We will return to non-linear-feedback shift registers in Section 5.7.

5.4 Random properties of shift register sequences

In the previous section we concluded that the maximum period of a non-linear-feedback shift register consisting of m sections is $2^m - 1$. Thus, as m increases, the period will also increase. The advantage of sequences with a large period is that their predictability is far smaller than that of sequences with a small period. However, this is not the only criterion for cryptographic applications. If, for instance, a certain text with a long period is represented by a sequence of first all zeros and then all ones, it will be totally unsuitable for enciphering, since the ciphertext will be equal to the plaintext as long as only zeros are processed. When ones are produced, the ciphertext will be equal to the inverted plaintext.

We must therefore search for other criteria for determining whether a sequence is sufficiently random for a given encipherment. Sufficiently random means that succeeding bits cannot be predicted easily if a given number of bits of the sequence is known. Golomb (1984) has stated three postulates which a sequence must satisfy in order to be 'sufficiently random'

Figure 5.6. An example of a non-linear-feedback shift register.

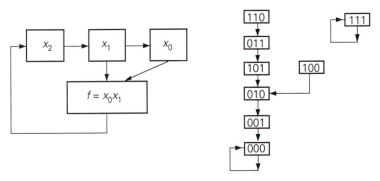

or, in other words, 'pseudorandom'. Before we can study these postulates, we must first introduce the concept of *autocorrelation*.

Suppose a sequence of bits is given by $X = (x_1, x_2, \ldots)$. From this, we can

Figure 5.7. De Bruyn graphs for $m = 1, \ldots, 5$.

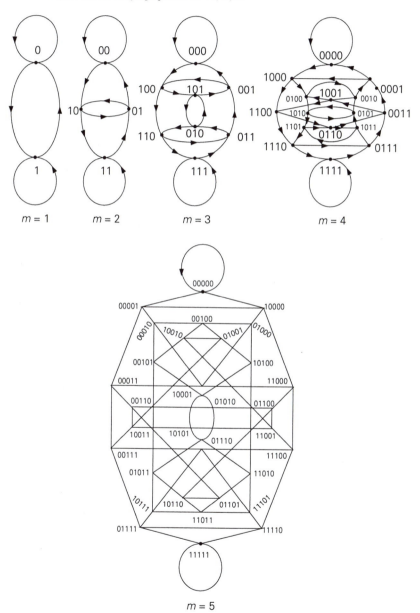

construct a new sequence which starts at a different element to x_1. The order of the elements remains the same. This results in a new sequence which is shifted with respect to the original sequence and can be described by:

$$x_i = x_{i+\tau},$$

in which τ is an integer equal to the number of places the sequence is shifted.

In general, the autocorrelation is defined as:

$$C(\tau) = \lim_{N \to \infty} \frac{1}{N} \sum_{i=1}^{N} x_i x_{i+\tau}. \tag{5.4}$$

If a given sequence has a period p, then the autocorrelation will be:

$$C_p(\tau) = \frac{1}{p} \sum_{i=1}^{p} x_i x_{i+\tau}. \tag{5.5}$$

However, there are many practical situations in which this formula cannot be applied directly. Assume we have a sequence consisting of a set of symbols (for instance, the letters of the alphabet); then we will have to assign values to the symbols to be able to calculate the autocorrelation. The problem then arises of how to choose these values.

Also, care must be taken in the case of binary sequences. Since binary multiplication is an asymmetrical operation, the autocorrelation will tend to zero. This problem can be solved by assigning the value $+1$ to the symbol 1 and the value -1 to the symbol 0.

Now, Golomb's three criteria for (pseudo)random sequences are:

(1) The difference between the number of zeros and the number of ones within one period must be as small as possible; often the number of ones is larger than the number of zeros.

(2) A *run* is defined as a series of identical symbols, preceded and followed by a different symbol. Within a period of the sequence, half of the runs will have a length equal to 1, a quarter of the runs will have a length of 2, an eighth of the runs a length equal to 3, etc. In other words, if $r(l)$ is the number of runs of length l and the total number of runs is equal to r, then provided $2^{-l}r > 1$:

$$r(l) = 2^{-l}r.$$

(3) When $\tau \neq p$, the autocorrelation will be a constant value. If $\tau = 0$ or $\tau = p$, the autocorrelation is equal to 1.

If a sequence satisfies each of these three points, it is called a pseudorandom sequence.

Example 1

The period p is equal to 15.

Sequence:	0 0 0	1	0 0	1 1	0	1	0	1 1 1 1
Values:	−1 −1 −1	1	−1 −1	1 1	−1	1	−1	1 1 1 1
Run lengths:	3	1	2	2		1	1	1 4

Postulate (1) has been satisfied: there are 8 ones and 7 zeros. As far as postulate (2) is concerned, we see that there are 8 runs, of which 4 have a length equal to 1, 2 have a length equal to 2 and one has a length equal to 3. We also find that one run has a length of 4. However, this does not contradict postulate (2), as for this case $2^{-l}r < 1$. Thus, postulate (2) has also been satisfied.

When $\tau = 0$, the autocorrelation is equal to 1. For all other values of τ, the autocorrelation will be equal to $-\frac{1}{15}$. We therefore find that postulate (3) also holds. Thus, we can conclude that the given sequence may be regarded as a pseudorandom sequence.   △

Example 2

The period p is equal to 15.

Sequence:	1 1 1	0	1	0 0 0	1	0	1 1 1	0 0
Values:	1 1 1	−1	1	−1 −1 −1	1	−1	1 1 1	−1 −1
Run lengths:	3	1	1	3	1	1	3	2

The number of ones and zeros corresponds to the requirements of postulate (1). The second postulate is not satisfied, however, as the number of runs of length 2 and length 3 differ from the stated requirements. When $\tau \neq p$, the autocorrelation is not constant, so postulate (3) does not apply. Therefore, this sequence cannot be regarded as a pseudorandom sequence.

                                                                                △

We have seen that a LFSR, comprising m sections, can generate sequences with a maximum period of $2^m - 1$. We will now investigate to what extent sequences with a maximum period can be regarded as pseudorandom sequences, by examining whether each of Golomb's postulates is satisfied.

Theorem 5.1

Consider a sequence with a maximum period of $2^m - 1$, generated by a LFSR. Denoting the number of ones and the number of zeros by $n(1)$ and $n(0)$ respectively, we find that:

$$n(1) - n(0) = 1. \tag{5.6}$$

Proof

A LFSR of m sections has a maximum of 2^m possible states. The last bit of the sequence representing each state appears at the output of the register. Therefore, the output sequence will contain as many zeros as ones. However, we have already remarked that the state $(00...0)$ must be disregarded, as in this state, the generated sequence does not have the maximum possible length. Thus, a LFSR will generate more ones than zeros. This can be expressed as:

$$n(1) = 2^{m-1} \quad \text{and} \quad n(0) = 2^{m-1} - 1$$

which results in:

$$n(1) - n(0) = 1. \qquad \square$$

Theorem 5.2

The number of runs of length l, generated by a LFSR with a maximum period of $2^m - 1$, is:

$$r(l) = 2^{-l}r, \tag{5.7}$$

in which $r(l)$ is the number of runs with a length equal to l and r is the total number of runs.

Proof

Three cases are distinguished, depending on the length of a run, i.e. $l = m$, $l = m - 1$ and runs whose length satisfies $0 < l < m - 1$. If $l = m$ the output sequence will consist of m zeros or m ones. The state $(00...0)$ is not permitted and therefore an output sequence of m zeros cannot occur. An output sequence of m ones, however, can occur, since the state $(11...1)$ is a valid state. Thus: $r(m) = 1$.

For the case in which $l = m - 1$, we find that $r(m - 1) = 1$. This is explained as follows. If an output sequence consisting of m ones is a valid output sequence, then there must be an output sequence of $(m + 2)$ bits which is equal to (011...10). This statement includes the states (011...1) and (11...10). A run of $(m - 1)$ ones is not possible. A run of $(m - 1)$ zeros, preceded by a one and followed by a one, can result from the states (10...0) and (0...01). Thus, we find that $r(m - 1) = 1$.

Finally, there are the cases for which $0 < l < m - 1$. Suppose we have a run of l ones. This implies that a zero must appear at the end of this run. Therefore, the number of remaining free positions for ones and zeros is equal to 2^{m-l-2}. This also applies to runs of l zeros, so again, there are 2^{m-l-2} possibilities. Hence:

$$r(l) = 2^{m-l-2} + 2^{m-l-2} = 2^{m-l-1}.$$

The total number of runs is r, so:

$$r = \sum_{l=1}^{m} r(l).$$

Substitution of the above expressions results in:

$$r = \sum_{l=1}^{m-2} r(l) + r(m - 1) + r(m)$$

$$= \sum_{l=1}^{m-2} 2^{m-l-1} + 2 = 2^{m-1}.$$

Since $r(l) = 2^{m-l-1}$ with $0 < l < m$ it immediately follows that:

$$r(l) = r \cdot 2^{-l}.$$

$\square$

Theorem 5.3

The autocorrelation of a sequence generated by a LFSR with a maximum period of $2^m - 1$ is:

$$C_p(\tau) = 1 \text{ when } \tau = p, \tag{5.8}$$

and

$$C_p(\tau) = -1/(2^m - 1) \text{ when } \tau \neq p. \tag{5.9}$$

Proof

If $\tau = p$ then eq. (5.5) can be rewritten as:

$$C_p(\tau) = \frac{1}{p}\sum_{i=1}^{p} x_i x_{i+p} = \frac{1}{p}\sum_{i=1}^{p} x_i^2 = \frac{p}{p} = 1.$$

The second half of the proof of Theorem 5.3 is based on the fact that the output sequences of a LFSR constitute a so-called Abelian group. This means that the result of each binary addition of two output sequences is a new output sequence itself. In Figure 5.8 the initial state (100) eventually results in an output sequence $g_1 = (0011101)$. The output g_2–g_7 sequences are generated by choosing a different initial state.

For the elements of the Abelian group $G = (g_1, g_2, ..., g_7)$, $g_i \in G$ and $g_j \in G$, it holds that $(g_i + g_j) \in G$. The calculation of $C(\tau)$, with $\tau \neq p$, is based on the values in the following table.

	binary representation								representation for autocorrelation						
output sequence x_i	0	0	1	1	1	0	1	x_i	−1	−1	1	1	1	−1	1
sequence $x_{i+\tau}$	+ 1	1	1	0	1	0	0	$x_{i+\tau}$	× 1	1	1	−1	1	−1	−1
sequence $x_i + x_{i+\tau}$	1	1	0	1	0	0	1	$x_i x_{i+\tau}$	−1	−1	1	−1	1	1	−1

When all the values of $x_i x_{i+\tau}$ are summed, we find

$$\sum_{i=1}^{p} x_i x_{i+\tau} = -1,$$

Figure 5.8. A LFSR consisting of three sections.

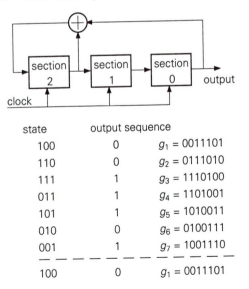

state	output	sequence
100	0	$g_1 = 0011101$
110	0	$g_2 = 0111010$
111	1	$g_3 = 1110100$
011	1	$g_4 = 1101001$
101	1	$g_5 = 1010011$
010	0	$g_6 = 0100111$
001	1	$g_7 = 1001110$
100	0	$g_1 = 0011101$

and therefore $c(\tau) = -\frac{1}{7} \cdot \sum_{i=1}^{p} x_i x_{i+\tau}$ is equal to the number of plus ones minus the number of minus ones.

Comparison with the binary representation reveals that it is also possible to calculate the difference between the number of zeros and the number of ones of the sum of the output sequence and shifted output sequence. As the output sequence and the shifted output sequence belong to an Abelian group, their sum will also be an element of an Abelian group. As a consequence, the number of zeros and ones in the sum of the output sequence and shifted output sequence is exactly equal to the number of zeros $n(0)$ and ones $n(1)$ in an arbitrary output sequence of the shift register. We have already calculated the values of $n(0)$ and $n(1)$ in the previous section. For the general case, we can now write ($\tau \neq p$):

$$C_p(\tau) = \frac{1}{p} \sum_{i=1}^{p} x_i x_{i+\tau} = \frac{1}{p} [n(0) - n(1)]$$

$$= \frac{1}{p} [(2^{m-1} - 1) - 2^{m-1}] = -\frac{1}{p} = -\frac{1}{2^m - 1}. \qquad \square$$

Hence, we can conclude that a sequence generated by a LFSR (consisting of m sections) with a maximum period of $2^m - 1$ will satisfy Golomb's postulates.

In addition to the three criteria for pseudorandom sequences introduced by Golomb, another criterion is often used, i.e. the so-called *complexity profile* (Rueppel, 1986). Let $X = (x_1, x_2, \ldots)$ represent a binary sequence. The linear complexity of the first n bits of the sequence is represented by $LC(n)$. The increase in $LC(n)$ as a function of n is called the linear complexity profile. A measure of the linear complexity is given by the minimum number of sections which is required for generating a certain subsequence. In Section 5.6 we will describe a method which can be used to find the minimum number of sections required, in order to generate a given subsequence. It can be demonstrated that the linear complexity profile of a truly random sequence is given by a straight line according to $n/2$, as illustrated in Figure 5.9. The extent to which the linear complexity profile of a given sequence deviates from this line is a measure of the pseudorandomness of the sequence.

5.5 The generating function

A very useful method of analysing LFSRs is one based on generating functions. In this section we will denote the sequence generated by a LFSR of m sections as $S = s_0, s_1,....$ With respect to S we can now define a so-called *generating function*. This function is a power series whose coefficients are equal to the terms of the sequence S, so:

$$G(x) = \sum_{i=0}^{\infty} s_i x^i. \tag{5.10}$$

This is a modulo 2 summation. Furthermore, the initial state is given by:

$$s_{-1}, s_{-2}, s_{-3}, \ldots, s_{-m}.$$

From this initial state, we can calculate the following elements of the sequence with the recursive expression of eq. (5.11):

$$s_i = \sum_{j=0}^{m-1} c_j s_{i-m+j}, \tag{5.11}$$

in which c_j is equal to either 0 or 1 and the summation is performed modulo 2.

This recursive formula clearly expresses the fact that each element is determined by the values of m preceding elements. It is, in fact, an alternative way of describing a LFSR. When $c_j = 1$, the contents of section j will influence the feedback function and when $c_j = 0$, the contents of the

Figure 5.9. Example of a linear complexity profile of the sequence (1000111101000011011111101000101).

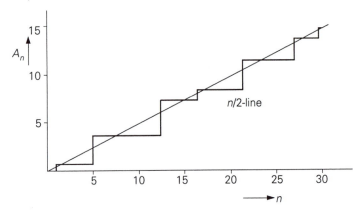

associated section will not influence the feedback function. Thus, the recursive expression of eq. (5.11) describes the shift register entirely.

By substituting the given expression, eq. (5.11), for s_i in eq. (5.10), the following formula is obtained:

$$G(x) = \sum_{i=0}^{\infty} s_i x^i = \sum_{i=0}^{\infty} \sum_{j=0}^{m-1} c_j s_{i-m+j} x^i$$

$$= \sum_{j=0}^{m-1} c_j x^{m-j} (s_{-m+j} x^{-m+j} + \ldots + s_{-1} x^{-1} + \ldots + s_i x^i + \ldots). \quad (5.12)$$

In this last expression the power series $G(x)$ reappears on the right-hand side of the equation as well. If this term is brought entirely to the left-hand side, the equation can be rewritten as:

$$G(x) = \sum_{j=0}^{m-1} c_j x^{m-j} (s_{-m+j} x^{-m+j} + \ldots + s_{-1} x^{-1})/f(x)$$

$$= s(x) / f(x), \quad (5.13)$$

in which $f(x)$ is given by:

$$f(x) = 1 - \sum_{j=0}^{m-1} c_j x^{m-j}. \quad (5.14)$$

The function $f(x)$ is a polynomial of the order m and is referred to as the *characteristic polynomial* of the sequence.

Assuming the initial state is given by (00000...1), then we can simplify the expression for $G(x)$ to

$$G(x) = c_0/f(x) = 1/f(x), \quad (5.15)$$

in which $c_0 = 1$. We have deliberately set this coefficient to 1, because otherwise, with $c_0 = 0$, the contents of section 0 would have no influence on the feedback function and the shift register of m sections would reduce to a simple $(m-1)$ section shift register.

The above analysis of the generating function has the advantage that it immediately reveals the period of the sequence generated by the shift register. A period p implies that:

$$s_i = s_{i+kp}, \text{ with } k = 0,1,2,\ldots . \quad (5.16)$$

Substitution of this equation in the expression for $G(x)$ results in:

$$G(x) = \sum_{i=0}^{\infty} s_i x^i = \sum_{k=0}^{\infty} \sum_{i=0}^{p-1} s_{i+kp} x^{i+kp}$$

$$= \sum_{i=0}^{p-1} s_i x^i \sum_{k=0}^{\infty} x^{kp}$$

$$= (s_0 + s_1 x + \ldots + s_{p-1} x^{p-1})(1 + x^p + x^{2p} + \ldots)$$

$$= (s_0 + s_1 x + \ldots + s_{p-1} x^{p-1})/(1 - x^p). \qquad (5.17)$$

The final step is achieved by dividing and multiplying by a factor $1/(1 - x^p)$.

By choosing the initial state $(00000...1)$, we have already seen that $G(x) = 1/f(x)$. Therefore, it holds that:

$$1/f(x) = (s_0 + s_1 x + \ldots + s_{p-1} x^{p-1})/(1 - x^p) \qquad (5.18)$$

and

$$f(x)(s_0 + s_1 x + \ldots + s_{p-1} x^{p-1}) = (1 - x^p). \qquad (5.19)$$

If the period is given by p, then the term $(1 - x^p)$ is a factor of the characteristic polynomial $f(x)$. The reverse also holds, i.e. if $f(x)$ is a factor of $(1 - x^p)$, then the sequence will have a period equal to p.

From this we can conclude that the smallest positive integer p, for which the characteristic polynomial is a factor of $(1 - x^p)$, must be equal to the period of the sequence. This value of p is also called the *exponent* of $f(x)$.

Previously, we saw that the length of a so-called maximum-length sequence generated by a shift register of m sections is $2^m - 1$. Therefore, in order to realise such a sequence it follows that the feedback should be such that the corresponding characteristic polynomial $f(x)$ is a factor of $(1 - x^{2^m-1})$.

Example

Assume $m = 3$ and $p = 2^m - 1 = 7$. The characteristic polynomial $f(x)$ must be a factor of $(1 - x^7)$. It can be demonstrated that:

$$1 - x^7 = (1 - x)(1 + x + x^3)(1 + x^2 + x^3).$$

A shift register consisting of three sections and with a feedback loop for which $f(x)$ is given by

$$f(x) = 1 + x^2 + x^3,$$

will generate a maximum-length sequence. From this characteristic polynomial it follows that $c_0 = c_1 = 1$ and $c_2 = 0$; this corresponds to the shift register depicted in Figure 5.5.   △

One may wonder how many possibilities there are for linear shift registers, which can generate maximum-length sequences. Whether a shift register is capable of producing maximum-length sequences depends on whether $f(x)$ is a *primitive polynomial*. A primitive polynomial of order m is an irreducible polynomial, which cannot be factorised further and which has an exponent equal to $2^m - 1$.

If $f(x)$ is a primitive polynomial of order m, then the associated shift register will produce a maximum-length sequence with a period equal to $2^m - 1$. This is easily explained by the fact that if $f(x)$ is a primitive polynomial of *the* order m, the exponent is equal to $2^m - 1$. Therefore, $f(x)$ will be a factor of $(1 - x^{2^m-1})$. Consequently, the period p is equal to $p = 2^m - 1$, which agrees with the case of maximum-length sequences.

The reverse also holds: when a shift register generates maximum-length sequences, then $f(x)$ must be a primitive polynomial. Suppose this were not true and that we could rewrite $f(x)$ as $f(x) = g(x)h(x)$. Denoting the order of $g(x)$ by m_g and the order of $h(x)$ by m_h, then it follows that $m_g + m_h = m$. It is clear that we should always be able to find an $a(x)$ and a $b(x)$ for which:

$$\frac{1}{f(x)} = \frac{a(x)}{g(x)} + \frac{b(x)}{h(x)}.$$

The terms $a(x)/g(x)$ and $b(x)/h(x)$ can each be associated with a shift register. The sequences generated with the aid of $a(x)/g(x)$ will have a maximum period of $2^{m_g} - 1$; those generated with $b(x)/h(x)$ will have a maximum period of $2^{m_h} - 1$. Therefore, the combination of these two shift registers will produce sequences with a period equal to the smallest common multiple of the two periods:

$$\text{period}[1/f(x)] = \text{period}[a(x)/g(x) + b(x)/h(x)]$$

$$\leq (2^{m_g} - 1)(2^{m_h} - 1) = 2^{m_g+m_h} - 2^{m_g} - 2^{m_h} + 1$$

$$= 2^m - 2^{m_g} - 2^{m_h} + 1 \leq 2^m - 3.$$

We assumed, however, that $f(x)$ produced maximum-length sequences, so:

$$\text{period}[1/f(x)] = 2^m - 1.$$

Since the period cannot be smaller than $2^m - 3$ and at the same time be equal to $2^m - 1$, we must conclude that the hypothesis $f(x) = g(x)h(x)$ was incorrect. Thus, $f(x)$ cannot be factorised further and is an irreducible and primitive polynomial.

The above has demonstrated that the requirement of a primitive (and irreducible) $f(x)$ is a necessary and sufficient condition for generating maximum-length sequences. Now, in order to answer the question of the required number of linear shift registers of length m, which generate maximum-length sequences, we must determine the number of primitive polynomials of the order m. Without proof, it is sufficient to state that it can be demonstrated (Golomb, 1984) that the number of primitive polynomials of order m is:

$$\varphi(2^m - 1)/m, \tag{5.20}$$

in which $\varphi(.)$ is Euler's totient function. *Euler's totient function* $\varphi(n)$ is defined as the number of positive integers smaller than n and totitive to n. A number is totitive to n when it is smaller than n and relatively prime to n. Relatively prime implies that they have no common factors.

Example
$n = 8$: 1, 3, 5 and 7 are smaller than 8 and relatively prime to 8. It therefore follows that $\varphi(8) = 4$. $\triangle$

A more general definition of $\varphi(n)$ is given by:

$$\varphi(n) = \begin{cases} 1 & \text{if } n = 1 \\ \prod_i u_i^{v_i-1}(u_i - 1) & \text{if } n > 1 \end{cases} \tag{5.21}$$

in which u_i represents the prime numbers into which n can be factorised and v_i is the number of occurrences of u_i.

It follows that for a prime number p, $\varphi(p) = p - 1$. For the product $n = pq$ of two prime numbers it holds that:

$$\varphi(n) = \varphi(p)\varphi(q). \tag{5.22}$$

Example
$n = 90 \rightarrow k = 2^1 \times 3^2 \times 5^1 \rightarrow u_1 = 2,\ v_1 = 1,\ u_2 = 3,\ v_2 = 2,\ u_3 = 5,\ v_3 = 1 \rightarrow$
$\rightarrow \varphi(n) = 2^0 \times 1 \times 3^1 \times 2 \times 5^0 \times 4 = 24$.
$n = 8 \rightarrow n = 2^3 \rightarrow u_1 = 2, v_1 = 3 \rightarrow \varphi(n) = 2^2(2 - 1) = 4$. $\triangle$

We can now construct a table of $\varphi(2^m - 1)/m$ for different values of m, in which the number of primitive polynomials $f(x)$, which is equal to the number of LFSRs capable of generating maximum-length sequences, is given. For instance, with $m = 8$, this number is equal to 16; if $m = 23$, it is 356960.

5.6 Cryptanalysis of LFSRs

In Section 5.4 we established that LFSRs are capable of generating sequences which satisfy the three requirements of Golomb for pseudorandom sequences. However, we must still be cautious. Considering the linear complexity profile of the LFSR, we see a large difference from what we may expect for a linear complexity profile of a (pseudo)random sequence. The reason for this is that in the case of a shift register of m sections, a cryptanalyst will need only $2m$ bits of ciphertext in order to reconstruct the shift register, as was demonstrated by Massey (1969). This implies that even for sequences with a very large period, the remaining bits can easily be predicted once only a small number of bits of the sequence is known. For instance, with $m = 16$, only 32 bits are required in order to generate the remaining 65500 bits.

This means that the complexity profile of a LFSR will tend towards a horizontal line with time. If the sequence is predictable from a certain point onwards, the complexity will no longer increase. Before we proceed with a demonstration of a cryptanalytic attack, we will first introduce the following theorem.

Theorem 5.4

If a sequence $s_0, s_1, \ldots, s_{N-1}$ is generated by a LFSR of m sections which cannot generate a sequence $s_0, s_1, \ldots, s_{N-1}, s_N$ and the latter sequence is generated by a shift register of m' sections, then

$$m' \geq N + 1 - m. \qquad (5.23)$$

Proof

If $m \geq N$, the proof is trivial; therefore assume $m < N$. Let the coefficients of the two shift registers be denoted by $c_0, c_1, \ldots, c_{m-1}$ and $c'_0, c'_1, \ldots, c'_{m'-1}$, respectively. Assume also that $m' < N - m$. The first m bits of the output sequence of the LFSR will correspond to those of the initial state. The remaining output bits of a sequence of N bits are determined by:

$$\sum_{j=0}^{m-1} c_j s_{i-m+j} = s_i, \quad i = m, m+1, \ldots, N-1. \tag{5.24}$$

Since we introduced the condition that the shift register cannot generate s_N, it follows that:

$$\sum_{j=0}^{m-1} c_j s_{N-m+j} \neq s_N. \tag{5.25}$$

For the second shift register of m' sections, we find that:

$$\sum_{k=0}^{m'-1} c'_k s_{i-m'+k} = s_i, \quad i = m', m'+1, \ldots, N. \tag{5.26}$$

From these two statements it follows that:

$$\sum_{j=0}^{m-1} c_j s_{N-m+j} = \sum_{j=0}^{m-1} \sum_{k=0}^{m'-1} c'_k s_{N-m+j-m'+k}. \tag{5.27}$$

The substitution of the right-hand term is valid, since $\{s_{N-m}, s_{N-m+1}, \ldots, s_{N-1}\}$ is a subset of $\{s_{m'}, s_{m'+1}, \ldots, s_{N-1}\}$. Furthermore, $\{s_{N-m}, \ldots, s_{N-1}\}$ is a subset of $\{s_m, \ldots, s_{N-1}\}$, which means we can exchange the summations, resulting in:

$$\sum_{j=0}^{m-1} c_j s_{N-m+j} = \sum_{k=0}^{m'-1} c'_k \sum_{j=0}^{m-1} c_j s_{(N-m'+k)-m+j}$$

$$= \sum_{k=0}^{m'-1} c'_k s_{N-m'+k}$$

$$= s_N. \tag{5.28}$$

This last result contradicts the fact that the Nth bit cannot be generated by the shift register of m sections. Therefore, $m' > N - m$ and consequently $m' \geq N + 1 - m$. $\qquad\square$

Suppose cryptanalysts have retrieved a sequence $s_0, s_1, \ldots, s_{N-1}$ and that they assumed that this sequence was generated by a linear shift register of no more than $N/2$ sections ($m \leq n/2$). The shift register in question can now be reconstructed in the manner shown below, i.e. we can now determine the number of sections m and the coefficients $c_0, c_1, \ldots, c_{m-1}$.

Let S represent the sequence $s_0, s_1, s_2, \ldots$ which is produced by the shift register. The number of sections of the shortest shift register capable of generating $s_0, s_1, \ldots, s_{N-1}$ is denoted by $M_N(s)$. If $M_{N+1}(s)$ is equal to the number of sections of the shift register which generates $s_0, s_1, \ldots, s_N$, then from Theorem 5.4 it immediately follows that:

$$M_{N+1}(s) \geq N + 1 - M_N(s). \tag{5.29}$$

However, since always $M_{N+1}(s) \geq M_N(s)$, it also follows that:

$$M_{N+1}(s) \geq \max[M_N(s), N + 1 - M_N(s)]. \tag{5.30}$$

It can be demonstrated that for the case of a LFSR, in fact, the two sides of eq. (5.30) are always equal, so for all values of N:

$$M_{N+1}(s) = \max[M_N(s), N + 1 - M_N(s)]. \tag{5.31}$$

The following theorem will affirm that if a shift register comprises m sections, cryptanalysts will require only $2m$ bits, in order to reconstruct the shift register.

Theorem 5.5

Consider a linear shift register of m sections. Let $M_N(s)$ denote the number of sections of the shortest shift register which is capable of producing $s_0, s_1, \ldots, s_{N-1}$. Then, it holds that:

$$M_N(s) = M_{2m}(s) \text{ for all } N \geq 2m. \tag{5.32}$$

Proof

If a shift register of $M_N(s)$ sections can generate the sequence $s_0, s_1, \ldots, s_{N-1}, s_N$ as well as $s_0, s_1, \ldots, s_{N-1}$, then $M_{N+1}(s) = M_N(s)$. However, if the shift register does not generate $s_0, s_1, \ldots, s_N$, then it holds that:

$$M_{N+1}(s) = \max\{M_N(s), N + 1 - M_N(s)\}. \tag{5.33}$$

Therefore, the number of sections will only change if and only if

$$N + 1 - M_N(s) > M_N(s);$$

i.e. when

$$N \geq 2M_N(s).$$

It is evident that the value of $M_N(s)$ will not change from $N = 2m$ onwards. Suppose that $M_{N+1}(s) > M_N(s)$ for a given $N > 2m$, then

$$M_{N+1}(s) = N + 1 - M_N(s) \geq 2m + 1 - m = m + 1.$$

This contradicts the fact that the sequence is generated by a shift register consisting of m sections. Therefore, $M_N(s) = M_{2m}(s)$ for all $N \geq 2m$. □

This theorem leads to the conclusion that cryptanalysts can determine the minimum length of a shift register which generates the entire sequence, provided that $n \geq 2m$. Furthermore, the algorithm of Figure 5.10 enables them actually to reconstruct the shift register step by step. The variables are initialised in step 1. The algorithm will reconstruct the shift register as the output sequence is made available bit by bit. Each time the algorithm arrives at step 2 it has determined the number of sections m and the characteristic polynomial $f(x)$ of a temporary LFSR, which can generate the first N bits which have been considered so far. As soon as N is equal to the length n of the available sequence, the algorithm terminates (step 2 and 9). However, until this point is reached, the algorithm will attempt to generate the following bit, based on the temporarily constructed shift register (step 3). If this is successful ($d = 0$), no adaptations need be made to the shift register and the algorithm can continue to the following bit (step 8). However, if the outcome is unsuccessful ($d = 1$), a new shift register must be constructed. If $2m \leq N$, the number of sections must be increased and the characteristic polynomial must be adapted (step 6). If $2m > N$, the same number of sections will suffice and only the characteristic polynomial need be adapted (step 7). In steps 6 and 7, N' is equal to the value of N when the last increase in the number of sections m was made and $b(x)$ is the characteristic polynomial which corresponds to $s_0, s_1, \ldots, s_{N'-1}$. The function $f(x)$ is stored temporarily in the function $t(x)$.

The final value of m represents the number of sections of the shift register. The coefficients $c_0, c_1, \ldots, c_{m-1}$ determine which sections influence the feedback function and can be derived from the expression for the characteristic polynomial:

$$f(x) = 1 + \sum_{j=0}^{m-1} c_j x^{m-j}. \tag{5.34}$$

This unambiguously defines m and $c_0, c_1, \ldots, c_{m-1}$ for a LFSR with a minimum number of sections, which can generate a given sequence $s_0, s_1, \ldots, s_{N-1}$ ($N > 2m$). The above method is illustrated by the example shown in Figure 5.11(a). The minimum required number of sections is 4. The characteristic polynomial is $f(x) = 1 + x + x^4$, from which it follows that $c_0 = 1$, $c_1 = 0$, $c_2 = 0$ and $c_3 = 1$. The corresponding shift register is depicted in Figure 5.11(*b*).

Figure 5.10. Algorithm for determining the characteristic polynomial $f(x)$ of a LFSR, which can generate a sequence $s_0, s_1, ..., s_{N-1}$ with a minimum number of sections.

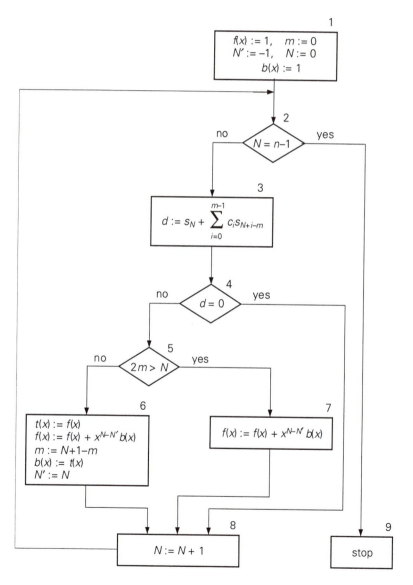

5.7 Non-linear shift registers

The fact that linear shift registers are very easily 'cracked' makes them unsatisfactory for most practical applications. Therefore, non-linear-feedback shift registers are frequently preferred. A disadvantage of non-linear-feedback shift registers from a designer's point of view is that, unlike linear shift registers, they cannot be adequately described in mathematical

Figure 5.11. Example of the cryptanalysis of a LFSR whose output sequence is (1010110010); (*a*) determination of the minimum number of sections and the characteristic polynomial; (*b*) reconstruction of the shift register.

step	$f(x)$	m	$b(x)$	N'	N	d
1	1	0	1	−1	0	
2, 3						1
4, 5, 6	$1 + x$	1	1	0		
8					1	
2, 3						1
4, 5, 7	1					
8					2	
2, 3						1
4, 5, 6	$1 + x^2$	2	1	2		
8					3	
2, 3						0
8					4	
2, 3						0
8					5	
2, 3						1
4, 5, 6	$1 + x^2 + x^3$	4	$1 + x^2$	5		
8					6	
2, 3						1
4, 5, 7	$1 + x + x^2$					
8					7	
2, 3						1
4, 5, 7	$1 + x + x^4$					
8					8	
2, 3						0
8					9	
2, 3						0
8					10	
2 STOP						

(*a*)

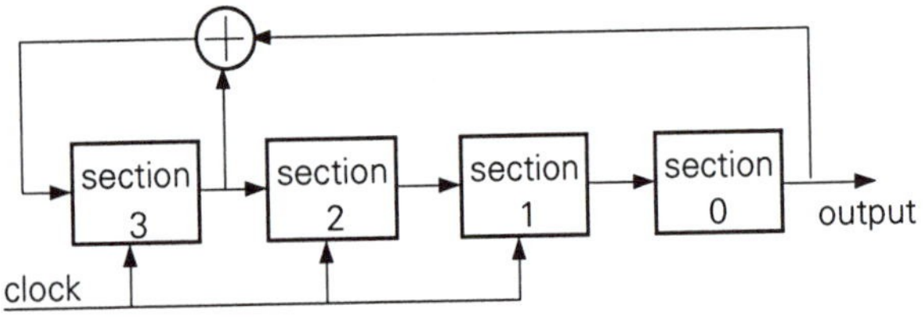

(*b*)

terms. On the other hand, this is precisely what gives the non-linear-feedback shift register its strength as a security measure.

We have already seen an example of a non-linear-feedback shift register in Figure 5.6. The most important differences from linear-feedback shift registers are that:

- the zero state, in which the sections of the shift register only contain zeros, can be followed by a non-zero state, thus allowing for a maximum period of 2^m;
- branches are allowed causing different states to be succeeded by the same state.

In Figure 5.5 a table was created, based on a given feedback function (in this example for a LFSR), which defines the successor of each state. Conversely, it is also possible to define the feedback function, based on a given table. Consider the following example.

Let x_0, x_1 and x_2 denote the contents of three shift register sections and $f(x_0,x_1,x_2)$ the non-linear feedback function we wish to determine. The values of the feedback function are given below as a function of the state:

x_2	x_1	x_0	$f(x_0,x_1,x_2)$
0	0	0	0
1	0	0	0
0	1	0	0
1	1	0	1
0	0	1	0
1	0	1	0
0	1	1	1
1	1	1	0

The corresponding function can be found as follows. For each row of the table in which $f(x_0,x_1,x_2) = 1$ we first calculate the product of x_0, x_1 and x_2, bearing in mind that when x_0, x_1 or x_2 is equal to 0, we use the value $x_0 + 1$, $x_1 + 1$ or $x_2 + 1$, respectively, instead.

The feedback function is found by adding the calculated products modulo 2. The values in the above table result in:

$$f(x_0,x_1,x_2) = (x_0 + 1)\, x_1 x_2 + x_0 x_1\, (x_2 + 1)$$

$$= x_0 x_1 x_2 + x_1 x_2 + x_0 x_1 x_2 + x_0 x_1$$

$$= x_0 x_1 + x_1 x_2.$$

The total number of possible feedback functions of a shift register of m sections is 2^{2^m}, as we saw in Section 5.3. However, these functions are not all equally suitable for a given application. If the state diagram contains branches, the maximum length of the generated output sequence will decrease. In the case of Figure 5.6, the shift register passes through each state (other than (000) and (111)) no more than once, after which it will only produce zeros, with a period of 1. Generally, branches must be avoided to allow for maximum length output sequences. There is a direct relation between the form of the feedback function and the presence of branches in the state diagram.

At a branch in the state diagram at least two different states will be succeeded by the same following state. If m is the length of the shift register, then these states will have $(m - 1)$ bits in common and only the bit in section 0 will differ. Consider, for instance, the states (10010) and (10011), which are both succeeded by (01001). The feedback function must be such that both (10010) and (10011) produce the value 0. In general terms, at a branch it must hold that:

$$f(x_{m-1},\ldots,x_1,x_0) = f(x_{m-1},\ldots,x_1,x_0 + 1). \tag{5.35}$$

This expression can be used to prove that there are no branches if and only if

$$f(x_{m-1},\ldots,x_1,x_0) = x_0 + f'(x_{m-1},\ldots,x_1)$$

In other words, it must be possible to rewrite $f(x_{m-1},\ldots,x_1,x_0)$ as the sum of x_0 and a new function $f'(x_{m-1},\ldots,x_1)$.

This is made clear by the following. Suppose

$$f(x_{m-1},\ldots,x_1,x_0) = x_0 + f'(x_{m-1},\ldots,x_1).$$

Then it immediately follows that also

$$f(x_{m-1},\ldots,x_1,x_0 + 1) = x_0 + 1 + f'(x_{m-1},\ldots,x_1).$$

Subtraction of these two equations results in:

$$f(x_{m-1},\ldots,x_1,x_0) - f(x_{m-1},\ldots,x_1,x_0 + 1) = 1$$

and therefore:

$$f(x_{m-1},\ldots,x_1,x_0) \neq f(x_{m-1},\ldots,x_1,x_0 + 1).$$

This means that there are no branches. This requirement is not merely a sufficient requirement for establishing the presence of branches, but it is also a necessary requirement, since the hypothesis that there are no branches implies that x_0 and $(x_{m-1},\ldots,x_1)$ are linearly independent. Therefore, $f(x_{m-1},\ldots,x_1,x_0)$ must be of the form $x_0 + f'(x_{m-1},\ldots,x_1)$.

Non-linear shift registers can be designed by defining non-linear feedback functions. However, it may also be possible to base the design on LFSRs. Unfortunately, though, if the design relies merely on the logical addition and logical multiplication of the output sequences of LFSRs it will not produce the desired result, as will become apparent below.

Logical addition

Suppose we have a linear-feedback shift register SR1 which generates a sequence $G_1(x) = S_1(x)/f_1(x)$ and a second linear-feedback shift register SR2 which generates a sequence $G_2(x) = S_2(x)/f_2(x)$. The new sequence which results from the binary addition of the output sequences of SR1 and SR2 can be described by the generating function:

$$G(x) = G_1(x) + G_2(x)$$

$$= s_1(x)/f_1(x) + s_2(x)/f_2(x)$$

$$= [s_1(x)f_2(x) + s_2(x)f_1(x)]/f_1(x)f_2(x). \qquad (5.36)$$

This is equivalent to a LFSR with a characteristic polynomial $f_1(x)f_2(x)$. In other words, a simple logical addition of the output sequences of two LFSRs does not result in non-linear sequences.

Logical multiplication

Multiplication of the two output sequences will result in a one in the new output sequence when the corresponding bits of the original output sequences are both equal to one; in all other cases, the output will be zero. This asymmetrical distribution of the ones and zeros after multiplication is evident when $n(1)$ and $n(0)$ are calculated. Consider two LFSRs with m_1 and m_2 sections, respectively, which both produce maximum-length sequences. The probability of a one in the new output sequence, $p(1)$, can be calculated from the probability of a one in the output sequence of a LFSR of m_i sections, which is equal to $2^{m_i-1}/(2^{m_i-1} - 1)$, $i = 1,2$. The new output sequence will only contain a 1 if both of the shift registers produce a one at the output and therefore $p(1)$ is equal to the product of the probability of a one in the output sequence of shift register 1 and the probability of a one in the output sequence of shift register 2.

Hence, we find that the probability of a one in the new output sequence is:

$$\frac{2^{m_1-1} \times 2^{m_2-1}}{(2^{m_1}-1)(2^{m_2}-1)} \approx \frac{2^{m_1-1} \times 2^{m_2-1}}{2^{m_1+m_2}} = \frac{1}{4}. \tag{5.37}$$

In other words, the number of zeros in the new output sequence is four times the number of ones. This does not satisfy Golomb's first postulate and therefore this method is not very suitable.

A better solution can be found by connecting the outputs of two feedback shift registers with a JK-flipflop.

JK-flipflop

Let x_i and y_i denote the bits of two LFSRs and w_i the output of a JK-flipflop. The function table of a JK-flipflop is given below:

x_i	y_i	$w_i = F(x_i, y_i)$
0	0	w_{i-1}
0	1	0
1	0	1
1	1	$w_{i-1} + 1$

If $x_i = y_i$, the JK-flipflop will produce a value w_i, which is equal to the previous bit w_{i-1}, or its inverse value ($w_{i-1} + 1$). The method described earlier enables us to find the function w_i. It follows that:

$$w_i = (x_i + 1)(y_i + 1)w_{i-1} + x_i(y_i + 1) + x_i y_i (w_{i-1} + 1)$$
$$= (y_i + x_i + 1)w_{i-1} + x_i.$$

The non-linearity of this output is evident from the first couple of terms of the output sequence:

$$w_1 = (x_1 + y_1 + 1)w_0 + x_1,$$

$$w_2 = (x_2 + y_2 + 1)w_1 + x_2$$
$$= (x_2 + y_2 + 1)(x_1 w_0 + y_1 w_0 + w_0 + x_1) + x_2$$

and so on.

However, this system is still not entirely secure. With a given output sequence it is possible to reconstruct half of the bits of the combined shift register. Consider the following values:

$$w_{i-1} = 0, w_i = 0 \Rightarrow x_i = 0,$$

$$w_{i-1} = 0, w_i = 1 \Rightarrow x_i = 1,$$

$$w_{i-1} = 1, w_i = 0 \Rightarrow y_i = 1,$$

$$w_{i-1} = 1, w_i = 1 \Rightarrow y_i = 0.$$

For instance, if the output sequence is (011100100110), then it will cost no effort to determine that the two LFSRs must have generated the sequences (??001??1??01) and (?1???01?01??).

The security of the system can be improved successfully with JK-flipflops, using the *Pless system* (1976) (see Figure 5.12). A total of eight LFSRs are connected by means of JK-flipflops, producing four bit streams. The output sequence is formed by periodically selecting a bit from these bit streams. Thus, the output bits in positions 1, 5, 9, ... etc. are supplied by SR1 and SR2, those in positions 2, 6, 10, ...etc. by SR3 and SR4, etc. In this manner, the final output sequence does not contain any successive bits of the JK-flipflops.

If each shift register generates maximum-length sequences and the lengths of the shift registers m_i, $i = 1,...,8$, are totitive, the period is:

$$p = \prod_{i=1}^{8} (2^{m_i} - 1).$$

The values for m_i in the system proposed by Pless are 5, 19, 7, 17, 9, 16, 11 and 13, which leads to a period p of $p \approx 1.519 \times 10^{29}$. This system can be used for simple applications with fairly light security requirements. However, it is not capable of withstanding a cryptanalytic attack with a powerful computer (see Siegenthaler (1985)).

Multiplexer

A better result is obtained by incorporating the shift registers in a so-called *multiplexer circuit* (see Figure 5.13). The output sequence is actually generated by SR2. The sequence generated by SR1 determines which bits of the output sequence of SR2 finally appear in the output sequence of the multiplexer.

The table provided overleaf is an example of the possible output of the multiplexer.

The selected address depends on the state of SR1 and is equal to the integer value of x_0 and x_1. The multiplexer will map these addresses onto a selected output stage. In the example shown the addresses (0,1,2,3) are mapped onto (3,2,0,1). This means that when the address is zero, bit x_3 of SR2 is output, and when the address is one, bit x_2 is selected, etc.

It is evident that when both shift registers are in their initial states simultaneously, the output sequence is repeated. Therefore the period will be no larger than the product of the separate periods.

SR1			address	stage	SR2				output sequence
x_0	x_1	x_2			x_0	x_1	x_2	x_3	
1	0	0	2	0	1	0	0	0	1
0	0	1	0	3	0	0	0	1	1
0	1	0	1	2	0	0	1	1	1
1	0	1	2	0	0	1	1	1	0
0	1	1	1	2	1	1	1	1	1
1	1	1	3	1	1	1	1	0	1
1	1	0	3	1	1	1	0	1	1

Figure 5.12. The Pless system.

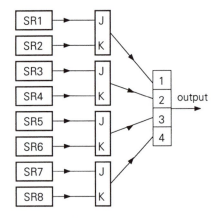

Figure 5.13. Connection of linear shift registers with a multiplexer circuit.

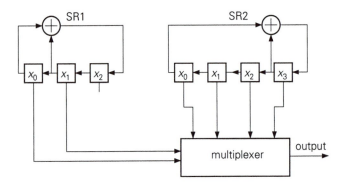

6

Public key systems

6.1 Introduction

In the previous chapters we examined cryptographic methods for concealing information which are based on a single key. This key is used by the transmitter for encrypting the information and by the receiver for deciphering the received massage. An important problem here is how to ensure the same secret key is available to both the transmitter and receiver. The classical solution is to employ couriers, who travel between the transmitter and receiver, as is often the case in diplomatic and military circles. Another possibility is to send the key along a different transmission line than the ciphertext. In all cases though, the key must be kept secret during transmission, since a person who possesses the key is able to decipher the text.

In a so-called *public key system* two keys are used: one for enciphering and one for deciphering a text. The cipher key is made public; anyone has access to this key and is permitted to use it for enciphering a text. A public key system is based on the principle that anyone is able to encipher a text, but only a few can decrypt the ciphertext. The system ensures that it is almost impossible to derive the decipher key from the cipher key. A system which relies on both a secret key and a public key is often referred to as an *asymmetrical cipher system*. A system based on a single (secret) key, such as the DES, is called a *symmetrical cipher system*.

Two important aspects of public key systems are the so-called *one-way functions* and the '*trapdoor*' *functions*. The term one-way function is applied to functions which are relatively easy to calculate themselves, but whose inverse function is far more difficult to compute. An example of this is the power function. It is relatively easy to determine the product of a number of equal factors. However, the reverse operation, i.e. finding the root of a quantity, is considerably more complicated. Trapdoor functions are, in fact, a variation on one-way functions; determining their inverse functions is extremely difficult, unless additional information is available.

6.2 The RSA system

One of the most well-known and popular public key systems is the *RSA system*, named after the first letters of the surnames of its designers (R. L. Rivest, A. Shamir and L. Adleman of the Massachusetts Institute of Technology (MIT), see Rivest *et al.* (1978)). The RSA system is based on the fact that it is relatively easy to calculate the product of two prime numbers, but that determining the original prime numbers, given the product, is far more complicated.

The RSA outline

First, two prime numbers are generated and their product is calculated and denoted by n: $n = pq$. Then a number e is determined, which satisfies the following expression:

$$3 < e < (p-1)(q-1) \tag{6.1}$$

and which is also relatively prime to $(p - 1)(q - 1)$. In other words, the largest common factor of e and $(p - 1)(q - 1)$ is 1. Finally, the value of e is used to determine another number, d, for which:

$$ed = 1 \ (\text{mod} \ (p-1)(q-1)). \tag{6.2}$$

The public key consists of the pair (e,n); the other values are kept secret. The encipherment of a text is performed as follows. A binary representation of the text is used, which is divided into blocks denoted by M. The cipher block C is computed by raising the decimal value of M to the power e and taking the remainder of a division by n:

$$C = M^e \ (\text{mod} \ n). \tag{6.3}$$

The ciphertext is deciphered in a similar manner, with the exception that d is used instead of e:

$$M = C^d \ (\text{mod} \ n). \tag{6.4}$$

●

The security of this system relies on the fact that it is almost impossible to calculate the value of d if only the public key (e,n) is known. In order to find d, the values of p and q must be known, too (see Section 6.2)). Since only n is publicly available, a cryptanalyst must determine p and q from this. If n is of the order of 200 decimal digits, this would take approximately 30 million years, with the current technology. Thus, the person who issues the public key (e,n) is the only person who knows the precise value of d and therefore also the only person able to decipher encrypted texts.

The example given below will demonstrate exactly how this cipher system works.

Example

Suppose two prime numbers are chosen: $p = 3$ and $q = 17$ (in practice, obviously large prime number are used; these values of p and q are only used as an illustration). Their product n is 51 and $(p-1)(q-1) = 32$.

We must now find a number e, between 3 and 32, which has no factor in common with 32. Let, for instance, $e = 7$. Then we can determine d, for which it must hold $ed = 1 \pmod{(p-1)(q-1)}$. It follows from this that $d = 23$, since $ed = 7 \times 23 = 161 = 1 \pmod{32}$.

The public key is given by the values $(7,51)$. If the message ready for encipherment is represented by $M = 2$, then calculation of the ciphertext yields:

$$C = M^e \pmod{n} = 2^7 \pmod{51} = 26.$$

Decipherment of the text requires knowledge of the values of d and n. Then it follows that:

$$M = C^d \bmod n = 26^{23} \pmod{51}$$

$$= 26^1 \, 26^2 \, 26^4 \, 26^{16} \pmod{51}$$

$$= 26 \times 13 \times 16 \times 1 \pmod{51} = 2,$$

which is the original message. A potential cryptanalyst will only have access to the key $(7,51)$. However, in order to be able to decipher the ciphertext, the cryptanalyst must somehow discover that $d = 23$. This can only be calculated if the values 3 and 17, which correspond to the key value 51, can be found. Here, this is rather simple; given the value 51, p and q can very quickly be determined. However, if the value of n is represented by some 200 bits, i.e. n is of the order 2^{100}, this will be almost impossible. △

In the above example, the calculation of 26^{23} (mod 51) is simplified by the following two characteristics of modulo arithmetic:

(i) $a = b$ (mod n) $\Rightarrow$ $a^2 = b^2$ (mod n).   (6.5)

(ii) ab (mod n) = $[a$ (mod n)] $\cdot$ $[b$ (mod n)].   (6.6)

The first characteristic is evident: $a = b$ (mod n) implies that a is equal to b plus a multiple of n: $a = b + kn$. Therefore, $a^2 = b^2 + (kn)^2 + 2bkn$ and it immediately follows that $a^2 = b^2$ (mod n).

The second characteristic can also be derived easily:

$$[a \text{ (mod } n)] \cdot [b \text{ (mod } n)] = (a + kn)(b + k'n)$$

$$= ab + ak'n + bkn + kk'n^2 = ab \text{ (mod } n).$$

Due to the second characteristic, the calculation of 26^{23} (mod 51) can be split into the calculation of lower powers of 26 modulo 51. These separate powers can then be computed with the aid of characteristic (i), which implies that if $26^2 = 13$ (mod 51) then $26^4 = 13^2$ (mod 51) etc. Thus, these two characteristics can reduce the complexity of the calculations considerably.

We have already remarked that the values for d and e are such that we can always find the inverse operation of a given encipherment or decipherment. This can also be concluded from the previous example. In order to demonstrate the existence of the decipherment corresponding to a given encipherment, or vice versa, we must introduce a theorem from the theory of numbers.

In Chapter 5 we introduced Euler's totient function $\varphi(n)$, which represents the number of positive integers smaller than n and relatively prime to n. Since for a given prime number p, $\varphi(p) = p - 1$, it follows for the product of two prime numbers p and q that:

$$\varphi(n) = \varphi(p)\,\varphi(q) = (p - 1)(q - 1) = n - p - q + 1. \qquad (6.7)$$

Euler's theorem, which forms the basis of the RSA system, states the following.

Theorem 6.1 (Euler's theorem)

For all a and n, which are relatively prime with respect to each other and for which $n > 0$ and $0 < a < n$, it holds that:

$$a^{\varphi(n)} = 1 \text{ (mod } n). \qquad (6.8)$$

Proof

For a given n, $a \in Z'_n$, with $Z'_n = (r_1,...,r_m)$ representing the set of all positive integers between 0 and n which are relatively prime to n. It is evident that by definition:

$$|Z'_n| = m = \varphi(n).$$

For all values of i, there must be a value of j ($1 \le i, j \le m$) such that

$$ar_i = r_j \,(\text{mod } n).$$

This is explained as follows. Since a and r_i are relatively prime to n, the product ar_i must also be relatively prime to n. The set Z'_n includes all the numbers which are relatively prime to n, so therefore $ar_i \,(\text{mod } n) \in Z'_n$. In other words, $ar_i \,(\text{mod } n)$ is identical to one of the elements r_j of Z'_n. For every i, there is exactly one j such that $ar_i = r_j \,(\text{mod } n)$.

It follows that

$$ar_1 ar_2...ar_m = r_1 r_2 ... r_m \,(\text{mod } n)$$

or

$$a^m(r_1 r_2 ... r_m) = (r_1 r_2 ... r_m) \,(\text{mod } n),$$

resulting in

$$(a^m - 1)(r_1 r_2 ... r_m) = 0 \,(\text{mod } n).$$

Since $(r_1 r_2 ... r_m)$ are relatively prime to n, we find that

$$a^m - 1 = 0 \,(\text{mod } n) \;\Rightarrow\; a^m = 1 \,(\text{mod } n).$$

Theorem 6.1 follows immediately by substituting $m = \varphi(n)$ in this last statement. □

Example

Let $n = 17$ and $a = 2$. Then $\varphi(17) = 16$. Euler's theorem states that $2^{16} = 1 \,(\text{mod } 17)$. Straightforward calculation confirms that this is correct, $2^{16} = 65536 = 1 + 3855 \times 17$. △

Euler's theorem can be used to prove that encipherment and decipherment using the RSA system are the inverse operation of each other.

Theorem 6.2

Consider a message M, which is enciphered according to the RSA system, resulting in a ciphertext C:

$$C = M^e \bmod n. \tag{6.9}$$

The receiver deciphers this message into

$$M' = C^d \bmod n, \tag{6.10}$$

ensuring that $ed = 1 \pmod{(p-1)(q-1)}$. Then for all cases:

$$M' = M. \tag{6.11}$$

Proof

From $\varphi(n) = (p-1)(q-1)$ it follows that:

$$ed = 1 \ (\bmod \ (p-1)(q-1)) = 1 \ (\bmod \ \varphi(n))$$

and therefore

$$ed = k\varphi(n) + 1. \tag{6.12}$$

The received ciphertext C will satisfy:

$$C = M^e \pmod{n}.$$

Using the value of d to decipher this text, we can write:

$$M' = C^d \pmod{n} = M^{ed} \bmod n. \tag{6.13}$$

With the aid of eq. (6.12) this can be rewritten as:

$$M' = M^{ed} \pmod{n} = M^{k\varphi(n)+1} \pmod{n}. \tag{6.14}$$

Euler's theorem now enables us to simplify this expression. Assuming M and n are relatively prime, it holds that:

$$M^{\varphi(n)} = 1 \pmod{n}, \tag{6.15}$$

which finally results in:

$$M' = M^{ed} = M^{k\varphi(n)+1} = (M^{\varphi(n)})^k \, M$$

$$= (1)^k \, M \pmod{n} = M \pmod{n}. \tag{6.16}$$

If M and n are not relatively prime, then they must have a common factor, either p or q. Supposing p is the common factor, then M and $M^{\varphi(p)}$ are relatively prime with respect to q. According to Euler's theorem, we can write:

$$(M^{\varphi(p)})^{\varphi(q)} = 1 \ (\text{mod } q)$$

or

$$M^{\varphi(n)} = 1 \ (\text{mod } q),$$

and also

$$M^{k\varphi(n)+1} = M \ (\text{mod } q). \tag{6.17}$$

Since we have just assumed that M contains a factor p, then it must also hold that

$$M^{k\varphi(n)+1} = M \ (\text{mod } p). \tag{6.18}$$

By combining these last two expressions we find that:

$$M' = M^{ed} = M^{k\varphi(n)+1} = M \ (\text{mod } n). \tag{6.19}$$

Of course the same derivation can be given for the case in which the common factor of M and n is equal to q, instead of p.   □

As mentioned earlier, the strength of the RSA system relies on the fact that is impossible to derive the values of p and q from n, provided these prime values are sufficiently large. However, this also means that extremely large numbers must be used in the calculations for encipherment and decipherment .

Example

Let $p = 47$ and $q = 59$, then $n = pq = 2773$ and $(p - 1)(q - 1) = 2668$. The value of e must be chosen somewhere between 3 and 2668. Assume $e = 17$. We can then calculate the value for d with $ed = 1 \ (\text{mod } (p - 1)(q - 1))$, or $17d = 1 \ \text{mod } 2668$. This results in $d = 157$. Assume further that the alphabet is represented by decimal values, i.e. a = 01, b = 02, c = 03, etc. and a blank space is given the value 00. The plaintext ready for encipherment is given as:

$$M = \text{MAKE CONTACT IMMEDIATELY}$$

or in decimal representation by:

$$M = 1301\ 1105\ 0315\ 1420\ 0103\ 2009\ 1313\ 0504\ 0901\ 2005\ 1225.$$

The message is enciphered by regarding a block of four digits as an individual message, which is encrypted separately: $M_1 = 1301$, $M_2 = 1105$

etc. Encipherment of M_1 results in: 1301^{17} (mod 2773) = xxxx. In the same manner we can encrypt the remaining messages:

$$C = \text{xxxx xxxx xxxx xxxx xxxx xxxx xxxx xxxx xxxx xxxx xxxx}.$$

Decipherment is as follows: C_1 = xxxx, and therefore $M_1 = \text{xxxx}^{157}$ (mod 2773) = 1301 etc. △

This method of encryption involves complex calculations, which impose heavy requirements as far as computational power and speed are concerned. The last example demonstrated that the calculation of d from $ed = 1$ (mod $(p-1)(q-1)$), for given values of e, p and q, and the computation of terms such as $X = M^e$ (mod n) is not easy, especially when large values are concerned. We will therefore present two algorithms with which these calculations can be simplified considerably.

The calculation of d from the expression $ed = 1$ (mod $\varphi(n)$) is facilitated by a variation of the so-called *Euclidean algorithm*. Suppose $r(0) = \varphi(n)$ and $r(1) = e$. What we, in fact, wish to accomplish is to calculate the parameters $r(2)$, $r(3)$, ..., $r(k)$ recursively, until $r(k) = 1$. At this point the calculations are terminated and the value for d is found directly. The calculation of $r(2)$, $r(3)$, etc. is performed as shown below:

$$r(0) \quad = \quad a(1) \quad \cdot r(1) \quad + r(2)$$

$$r(1) \quad = \quad a(2) \quad \cdot r(2) \quad + r(3)$$

$$r(2) \quad = \quad a(3) \quad \cdot r(3) \quad + r(4)$$
$$\vdots \qquad \quad \vdots \qquad \vdots \qquad \vdots$$
$$r(k-2) \quad = \quad a(k-1) \cdot r(k-1) + r(k).$$

The values of $a(1)$, $a(2)$ etc. are chosen so that for all k, $r(k) < r(k-1)$. The above statement now enables us to express $r(2)$, $r(3)$, ..., $r(k)$ in terms of $r(0)$ and $r(1)$ alone:

$$r(2) = r(0) - a(1) \cdot r(1)$$

$$r(3) = r(1) - a(2) \cdot r(2) = -a(2) \cdot r(0) + \left[1 + a(1)a(2)\right] r(1)$$

$$\vdots$$

This process is terminated as soon as $r(k) = 1$. This value of $r(k)$ can also be expressed in terms of $r(0)$ and $r(1)$ alone. Suppose we have found $r(k) = u \cdot r(0) + v \cdot r(1)$, then it is immediately evident that v must be equal to the value of d. Since $r(k) = 1$ implies that $1 = u \cdot r(0) + v \cdot r(1) = u \cdot \varphi(n) + v \cdot e$, it follows that $ve = 1 \pmod{\varphi(n)}$ and therefore that v is equal to d.

Example
Determine the value of d, given $7d = 1 \bmod 32$.

Suppose $r(0) = 32$ and $r(1) = 7$. Substitution in the recursive expressions yields:

$$32 = 4 \times 7 + 4 \Rightarrow r(2) = r(0) - 4\,r(1)$$

$$7 = 1 \times 4 + 3 \Rightarrow r(3) \ = r(1) - r(2) = -\,r(0) + 5\,r(1)$$

$$4 = 1 \times 3 + 1 \Rightarrow r(4) \ = r(2) - r(3) = r(0) - 4\,r(1) - (-r(0) + 5\,r(1))$$

$$= 2\,r(0) - 9\,r(1),$$

from which it follows that $d = -9 = 23 \bmod 32$. △

At the beginning of this section we introduced a method for calculating the value of a power modulo n. This method for calculating $X = M^e \pmod{n}$ can be described in more general terms by the following:

(1) Write e in a binary form, so: $e = e_k, e_{k-1}, \ldots, e_1, e_0$.
(2) Set $X := 1$.
(3) Let for $i = k, k-1, \ldots, 0$:
    (*a*) $X :=$ remainder of (X^2/n),
    (*b*) if $e_i = 1$ then $X :=$ remainder of (XM/n).
(4) If $i = 0$ then stop. The current value of X is the required value.

Example
Calculate $X = 2^5 \pmod{11}$. We can write $M = 2$, $n = 11$, $e = 5$ or $e = 101$ in binary form.

Execution of the above algorithm yields the following results:

$k = 2$: $X :=$ remainder of $(1/11) = 1$
        since $e_2 = 1$: $X :=$ remainder of $(1 \times 2/11) = 2$;

$k = 1$: $X :=$ remainder of $(2^2/11) = 4$;

$k = 0$: $X :=$ remainder of $(4^2/11) = 5$
        since $e_0 = 1$: $X :=$ remainder of $(5 \times 2/11) = 10$.

Thus we find that $X = 2^5 \pmod{11} = 10 \pmod{11}$. △

The RSA system relies on very large prime numbers and therefore it is worthwhile to investigate methods of generating large prime numbers easily. The generation of sufficiently large prime numbers is a problem in its own right and can cost a considerable amount of calculating time. The best solution is to generate numbers with a random generator and to check whether each number is prime or not. This may seem a very impractical method, since the fraction of numbers which are prime is rather small. However, the absolute number of prime numbers available is large and therefore testing randomly generated numbers is a sensible method. It is not feasible to test whether a given number is prime or not by searching for possible factors; this would require too much time, which is precisely the characteristic on which the security of the RSA system is based. However, there are certain tests which can provide a fairly definite answer to this question. One of these tests is the primality test, as proposed by Solovay and Strassen (1977/8).

Let a candidate prime number be denoted by p and consider a value of a, selected from $a \in (1, ..., p-1)$. The largest common divisor of a and p is denoted by LCD(a,p). If LCD(a,p) $\neq$ 1, then p is not a prime number. If LCD(a,p) = 1 then check the following statement:

$$J(a,p) = a^{(p-1)/2} \pmod{p}, \tag{6.20}$$

in which $J(a,p)$ is the *Jacobi symbol*, which will be explained later.

If p is a prime number, the above statement will be true for all values of a. However, if p is not a prime number, then the above statement will be false in more than 50% of the cases. This implies that if the statement is true for 100 different values of a, the probability of p actually being a prime number will be larger than $1 - 2^{-100}$. In other words, as we verify the statement for all values of a and find that the statement is true each time, the probability of actually having found a prime number will increase. However, this method does not provide an absolute guarantee. It is therefore possible that very occasionally, a non-prime number will be used in the RSA system, without this being noticed. In this case, n can be written as the product of three, rather than two prime numbers. This may provide a potential weak spot for cryptanalysis.

The Jacobi symbol $J(a,p)$ represents a function which can assume one of only two values: either 1 or -1. For all values of p it is true that $J(1,p) = 1$. For other values of a and p, the Jacobi symbol can be calculated with the following expressions:

a even: $J(a,p) = J(a/2,p) \cdot (-1)^{(p^2-1)/8}$

a odd: $J(a,p) = J(p \pmod a),a)\cdot(-1)^{(a-1)(p-1)/4}$.

Example
Let $p = 17$. Choose a random value for a, say $a = 10$. It then holds that $LCD(10,17) = 1$. The value of $J(10,17)$ is:

$$J(10,17) = J(5,17)\cdot(-1)^{(17^2-1)/8} = J(5,17)\cdot(-1)^{36} = J(5,17)$$

$$= J(17 \pmod 5),5)\cdot(-1)^{(5-1)(17-1)/4} = J(2,5) = J(1,5)\cdot(-1)^{(5^2-1)/8}$$

$$= -1.$$

Calculation of $a^{(p-1)/2} \pmod p$ yields:

$$a^{(p-1)/2} \pmod p = 10^8 \pmod{17} = -1 \pmod{17}.$$

Thus, we have satisfied eq. (6.20). In this example, it is, in fact, trivial, because we already knew that 17 is a prime number. If p is large, the validity of eq. (6.20) only implies a large probability that p is a prime number. △

As well as probabilistic methods for determining whether a number is probably prime or not there are also deterministic methods which can actually prove this. However, these methods are usually far more complex and time-consuming than the probabilistic methods.

Not every number which passes the primality test is suitable for use in the RSA system. Therefore, we generally distinguish between weak and strong prime numbers. There are certain types of prime number for which it would be relatively easy to find the corresponding factorisation of n, if these were used for calculating the value of n. Therefore, in general, the following rules must be obeyed when selecting p and q:

(*a*) p must be chosen such that the factorisation of $p - 1$ has a large prime factor r.

(*b*) $p + 1$ must also have a large prime factor s.

(*c*) $r - 1$ must have a large prime factor t.

The same restrictions are imposed on q. Furthermore, p and q must be large, but their difference $|p - q|$ must also be large.

In practice, the following procedure can be followed for generating a prime number p:

– First, an arbitrary prime number s is generated, for instance by searching for the first prime number larger than a given starting value s_0, based on

the primality test of Solovay and Strassen. In the same manner a second
prime number t is generated.

- r is constructed from t. Since $r - 1$ must contain a prime factor t and r
  itself is odd, it must hold that $r = 1 \pmod{2t}$. Each value of $r = 2it + 1$ is
  tested on its primeness, for successive values of i, until a suitable value of
  r is found.
- Then a value for p can be determined, based on r and s, which satisfies
  criteria (*a*) and (*b*). This is accomplished as follows:
  - Calculate $u(r,s) = s^{r-1} - r^{s-1} \pmod{rs}$.
  - If $u(r,s)$ is odd, then let $p_0 = u(r,s)$; if not $p_0 = u(r,s) + rs$.
  - Calculate p using $p = p_0 + 2jrs$, for a given value of j, depending on
    the required order of magnitude of p and perform a primality test on p.
  - Take successive values of j and perform primality tests on p, until a
    prime number is found.

It will become evident that this procedure will result in a prime number p,
which satisfies conditions (*a*) and (*b*), when the following is considered.

In order for (*a*) and (*b*) to be satisfied, there must be a k and h such, that:

$$p = 2hr + 1 = 2ks - 1,$$

or

$$p = 1 \pmod{2r} = -1 \pmod{2s}.$$

With Euler's theorem this leads to:

$$s^{r-1} = 1 \pmod{r}.$$

It also holds that $r^{s-1} = 0 \pmod{r}$. This implies that

$$u(r,s) = s^{r-1} - r^{s-1} = 1 \pmod{r} = 1 + k'r.$$

If $u(r,s)$ is even, i.e. k' is odd, it follows that:

$$
\begin{aligned}
p = p_0 + 2jrs &= u(r,s) + rs + 2jrs \\
&= 1 + k'r + (2j + 1)rs \\
&= 1 \pmod{2r}.
\end{aligned}
$$

If $u(r,s)$ is odd, i.e. k' is even, it follows that

$$
\begin{aligned}
p = p_0 + 2krs &= u(r,s) + 2jrs \\
&= 1 + k'r + 2jrs \\
&= 1 \pmod{2r}.
\end{aligned}
$$

In the same manner it is possible to prove that $p = -1$ (mod $2s$). We can thus conclude that the value found for p is a strong prime number.

As far as the RSA system is concerned, there is no faster method of attack than factorisation. We must therefore consider the practical security, which depends on the computing power available for performing these complicated calculations. In 1988 Caron and Silverman managed to factorise a 90-digit number into two prime numbers of 41 and 49 digits, with the aid of 24 SUN-workstations (Caron and Silverman 1988). The required processing time was about six weeks (elapsed time). In the same year, Lenstra and Manasse successfully factorised a prime number of 96 digits (Lenstra and Manasse 1990). They employed a large number of computers, which were inter-connected by a combination of local area networks and electronic mail. The whole operation took 23 days, which effectively worked out to 10 years of CPU time. In 1990 they factorised a number of 138 digits in 50 days (Lenstra and Manasse 1991). Unfortunately, though, it is difficult to estimate the real value of this kind of incidental result.

At present, with the current technology it would still take approximately 500 years to factorise a random number of 130 digits, assuming the computer used is capable of performing 1 million operations per second. Obviously, this kind of astronomical investment is in no way compensated by the advantage a potential cryptanalyst may have by being able to factorise a large number. In practice, numbers of 768 bits (approximately 230 decimal digits) are recommended. Factorisation will then take at least 10^8 years using 1 million operations per second.

It is evident that despite the algorithms for reducing the total number of calculations, the RSA system still requires considerable computational power for processing such large numbers. For this reason, in practice, the RSA system is not especially well suited for real-time encryption of large amounts of data. With a key of 512 bits (154 digits), the processing speed of the RSA implemented in VLSI technology is no larger than 64 kbit/s. For the DES, hardware is being produced that can achieve 1 Gbit/s. The RSA system is therefore often used for enciphering limited amounts of data, for instance for the transportation of secret keys. The actual data can be encrypted with the DES algorithm and only the secret key of the DES algorithm is enciphered with the RSA system.

6.3 The knapsack system

A second public key system is based on the so-called knapsack problem. The knapsack problem can be described as follows. The vector

$$A = (a_1, a_2, \ldots, a_n)$$

consists of positive integers. The elements of this vector are multiplied by a binary vector, denoted by

$$X = (x_1, x_2, \ldots, x_n),$$

in which every x_i, $i = 1, \ldots, n$, is either 0 or 1. This results in the sum S:

$$S = \sum_{i=1}^{n} a_i x_i. \tag{6.21}$$

If X and A are given, the value of S can be calculated without any effort. However, if S and A are given, it is considerably more difficult to calculate X. This problem is known as the knapsack problem. See Figure 6.1. A knapsack is filled with a selection of objects from a large set; each object has a different weight (the elements of A correspond to the different weights of the objects). Is it possible to determine which items are in the knapsack, if the total weight, S, is known? In other words, we are looking for the elements of X, where a zero means that an item is not in the knapsack and a one means that it is. Experience has taught us that when A is sufficiently large (more than 100 elements) it is almost impossible to calculate X from a

Figure 6.1. The knapsack problem.

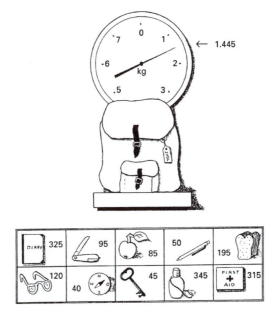

given S and A.

Now, the corresponding concept on which a cipher system can be based is the following. Suppose A is selected such that S can easily be calculated from X and A, but that the calculation of X given S and A is almost impossible, unless additional information is available. The vector X would then represent the plaintext, which could be converted to a ciphertext S with the aid of A. After the ciphertext has been received, it would be converted back to X.

Merkle and Hellmann (1978) have investigated what requirements are necessary for the vector A in order to be able find X easily from S. An example is given by the so-called *super-increasing sequence*. A super-increasing sequence is a sequence of numbers, whose elements are larger than the sum of the preceding elements. An example of such a sequence is:

$$A' = (141,203,427,981,2406).$$

Suppose S is 3590. Now, the knapsack problem can be solved easily. If $S = 3590$, x_5 must be equal to 1. The sum of the remaining elements of the sequence A' is smaller than 2406 and therefore cannot sum to 3590 without x_5. Thus, 2406 must contribute to S. The residue is $3590 - 2406 = 1184$. The same conclusion can be drawn with respect to x_4, so $x_4 = 1$. Now the remainder is $1184 - 981 = 203$. This eventually results in the vector $X = (0,1,0,1,1)$.

If super-increasing sequences are used, X can be retrieved from S. Thus, the principle of super-increasing sequences offers the possibility of unambiguous decipherment. However, before this principle can be applied in a crypto-system a method of concealing the super-increasing sequences must be found, to prevent unauthorised persons cracking the ciphertext. This can be done as follows.

Knapsack system with concealed super-increasing sequences

Two numbers, u and v are selected, which are relatively prime and for which also $u > \Sigma_i\, a_i$. The knapsack vector A, which is a super-increasing sequence, is transformed to a vector B, whose elements satisfy:

$$b_i = va_i \text{ (mod } u), \text{ for all values of } i. \tag{6.22}$$

The vector B is made public; u, v and A are kept secret. The vector X can now be enciphered to S, according to $S = BX$. The vector B will generally not be a super-increasing sequence and, consequently, anyone who does not have access to additional information is unable to decipher S. Decipherment of the ciphertext is only possible if u, v and A are available. Then the following procedure can be used. The inverse v^{-1} of the vector v can be computed from the expression:

$$v\,v^{-1} = 1 \ (\text{mod } u). \tag{6.23}$$

Once S has been received, it is multiplied by v^{-1} (mod u). This leads to

$$v^{-1} S \bmod u \ = v^{-1} \sum_{i=1}^{n} b_i x_i \ (\text{mod } u)$$

$$= v^{-1} \sum_{i=1}^{n} v\, a_i x_i \ (\text{mod } u)$$

$$= \sum_{i=1}^{n} v^{-1} v\, a_i x_i \ (\text{mod } u)$$

$$= \sum_{i=1}^{n} a_i x_i \ (\text{mod } u). \tag{6.24}$$

From this last term we can draw the conclusion that S can be reduced to the sum of the elements of a super-increasing vector and that, consequently, we can solve this expression for X.

Example

An example of a knapsack vector is $A = (3,5,9,19)$, which is a super-increasing sequence. Let $u = 40$ (then $u > 36$ is true) and $v = 7$. The inverse vector v^{-1} can be calculated as follows

$$v{\cdot}v^{-1} = 1 \ (\text{mod } u) \ \Rightarrow \ 7v^{-1} = 1 \ (\text{mod } 40) \ \Rightarrow \ v^{-1} = 23 \bmod 40.$$

The values of the elements of the enciphered sequence are calculated with $b_i = v{\cdot}a_i$ (mod u) $= 7a_i$ (mod 40). Then, the knapsack vector B will result in: $B = (21,35,23,13)$. Encipherment of, for instance, $X = (0110)$ produces: $S = BX = 35 + 23 = 58$.

After transmission, the receiver will multiply S by v^{-1}. X can then be found by applying the principle of super-increasing sequences, so $v^{-1} S$ (mod u) $= 23 \times 58$ (mod 40) $= 14$. The leads to $\sum_{i=1}^{4} a_i x_i = 14$ and finally to $X = (0110)$.

$\triangle$

6.4 Cracking the knapsack system

It was assumed for a long time that the knapsack method was entirely safe. However, let us consider the following situation. Suppose we are using a communication network, which consists of a centrally located main-frame, to which a large number of smaller systems are connected (for instance, host-terminal links). The communication between the host and the terminals is based on knapsack vectors; each connection has its own knapsack vector. At a certain point in time, the host wishes to send the same message X to all the terminals. The corresponding ciphertexts are:

$$S_k = \sum_{i=1}^{n} b_{ik} x_i, \tag{6.25}$$

where the index k is related to the address of the smaller systems. This is, in fact, a system of linear equations. The values of b_{ik} are known, since these are elements of the public key. Therefore, if the number of subsystems connected to the central system is larger than n, then by using straight-forward linear algebra, a cryptanalyst is able to find the plaintext X! Obviously, we must conclude that a single message should not be sent in more than one knapsack.

In 1984 Shamir presented a general method for cracking a knapsack system based on super-increasing sequences (Shamir 1984). In the previous section we have already seen that for all i it holds that:

$$b_i = va_i \; (\text{mod } u). \tag{6.26}$$

The reverse will also hold:

$$a_i = wb_i \; (\text{mod } u), \text{ for all } i, \tag{6.27}$$

in which $w = v^{-1} \; (\text{mod } u)$.

The super-increasing sequence A is secret, as well as w and u. In order to crack the algorithm, we must find a pair (w,u) for which the resulting a_is form a super-increasing sequence, whose sum is smaller than u. If these pairs can be found, then the message X can be retrieved from S. Shamir discovered a method for generating such pairs, which consists of two steps:

(1) Find a number of small intervals within [0,1], in which w/u must lie.
(2) These intervals can then be divided into subintervals, to enable a more direct search for a pair of whole numbers whose ratio is equal to w/u; then candidates for the super-increasing sequence A can be generated.

An attack can be outlined as follows. Suppose u is of the order of dn, with d a constant of proportionality. Since A is a super-increasing sequence, a_i will be of the order of:

$$a_1: dn - n \qquad \text{bits,}$$
$$a_2: dn - n + 1 \qquad \text{bits,}$$
$$\vdots \quad \vdots$$
$$a_i: dn - n + i - 1 \ \text{bits,}$$
$$\vdots \quad \vdots$$
$$a_n: dn - 1 \qquad \text{bits.}$$

In Figure 6.2, the curve wb_i (mod u) is plotted; we will refer to this curve as the b_i-curve. Each b_i-curve has exactly b_i zeros. The distance between two zeros is equal to u/b_i. The gradient of the teeth of the resulting sawtooth graph, with respect to the horizontal w-axis is $\tan(\alpha) = b_i$.

Consider the b_1-curve. It must hold that (see also eq. (6.27)):

$$a_1 = wb_1 \ (\text{mod } u). \tag{6.28}$$

The true value of w, as used by the algorithm, must satisfy eq. (6.28). The distance from this value of w to the closest zero to left, denoted by x_1, will be no larger than approximately 2^{-n}. This is explained as follows. The (mod u) operation in eq. (6.27) will ensure that all b_is, including b_1, will be of the same order of magnitude as u, i.e. dn bits which correspond to an order of magnitude of 2^{dn}. Since $\tan(\alpha) = b_1$, a_1 ($= wb_1$) is of the order of 2^{dn-n} and b_1 is of the order of 2^{dn}, the distance x_1 will be no larger than $2^{dn-n}/b_1 \approx 2^{-n}$. We can conclude from the distance between the zeros that w must lie close to a zero. However, we cannot predict which of the b_1 zeros

Figure 6.2. The b_i-curve as a function of w.

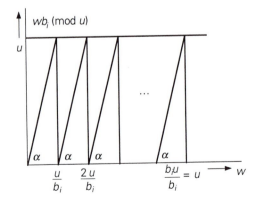

this is. Suppose it is the k_1th zero: $k_1 u/b$. Then it follows that:

$$\frac{k_1 u}{b_1} \le w \le \frac{k_1 u}{b_1} + 2^{-n}. \tag{6.29}$$

For the b_2-curve, we can state that w must lie within a distance no larger than $2^{dn-n+1}/b_2 \approx 2^{-n+1}$ from one of the zeros of the b_2-curve, since a_2 is of the order of 2^{dn-n+1}. Assuming this zero is denoted by k_2, it holds that:

$$\frac{k_2 u}{b_2} \le w \le \frac{k_2 u}{b_2} + 2^{-n+1}. \tag{6.30}$$

In general, we find for the b_i-curve:

$$\frac{k_i u}{b_i} \le w \le \frac{k_i u}{b_i} + 2^{-n+i-1}. \tag{6.31}$$

As w must lie close to a zero for all b_i-curves, it follows that all these zeros must lie close together. Therefore, we can approximate w by locating clusters of zeros along the horizontal axis.

This observation introduces the question of how many b_i-curves we must plot before we actually find a cluster of zeros with a given uncertainty. Apparently, finding just four zeros close together provides sufficient certainty that a cluster of zeros has been found.

Let us consider t b_i-curves. The next zero closest to $k_1 u/b_1$ on the b_1-curve must lie within the interval

$$\left[\frac{k_1 u}{b_1} - \frac{u}{2b_i}, \frac{k_1 u}{b_1} + \frac{u}{2b_i} \right]. \tag{6.32}$$

We will assume that the probability of a zero occurring at a given point in this interval is the same for all points of the interval. Regarding the distance between the zero k_1 of the b_1-curve and the closest zero of the b_i-curve, we find that the probability of this distance being smaller than 2^{-n+i-1} is:

$$2^{-n+i-1}/(u/b_i) = 2^{-n+i-1}.$$

The probabilities that the zeros of curves b_2–b_t lie close to zero k_1 of curve b_1 are given by:

$$2^{-n+1} \times 2^{-n+2} \dots 2^{-n+t-1} = 2^{-n(t-1)+t(t-1)/2}.$$

The value of w may lie close to any of the b_1 zeros and therefore, the probability P of a cluster of t zeros is:

$$P \approx b_1 \cdot 2^{-n(t-1)+t(t-1)/2} \approx 2^{dn-n(t-1)+t(t-1)/2}.$$

Suppose $d = 2$, $n = 100$ and $t = 4$, then it follows that $P \approx 2^{-94}$. This means that the probability of accidentally finding a cluster of four zeros is extremely small. We can therefore conclude that w will almost certainly lie close to a cluster of zeros of four b_i-curves.

The next step is to decide how to determine the value of u, as this is also unknown. Also, we must devise a method of finding a cluster.

We now introduce a new parameter $s = w/u$. Thus, the expression wb_i (mod u) can be rewritten as sb_i (mod 1). It is evident that the plot of sb_i (mod 1) will be identical to that of Figure 6.2. Consider Figure 6.3. This has the same number of zeros and the same gradient sawtooth as in Figure 6.2. The only difference is that the distance between the zeros is reduced by a factor 2^{dn} ($u \approx 2^{dn}$). Therefore, we can also use Figure 6.3 in our search for clusters of zeros.

In order to find a cluster of four zeros, we must solve three linear equations. From

$$\frac{k_i u}{b_i} \le w \le \frac{k_i u}{b_i} + 2^{-n+i-1}$$

it follows that:

$$\frac{k_i}{b_i} \le \frac{w}{u} \le \frac{k_i}{b_i} + 2^{-dn-n+i-1}. \tag{6.33}$$

Thus, the largest distance between (k_1/b_1) and (k_2/b_2) is no more than $2^{-dn-n+1}$. This results in:

Figure 6.3. The b_i-curve as a function of s.

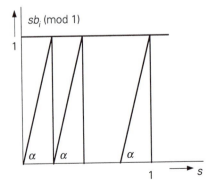

$$0 \leq \frac{k_1}{b_1} - \frac{k_2}{b_2} \leq 2^{-dn-n+1},$$

$$0 \leq \frac{k_1}{b_1} - \frac{k_3}{b_3} \leq 2^{-dn-n+2}, \qquad (6.34)$$

$$0 \leq \frac{k_1}{b_1} - \frac{k_4}{b_4} \leq 2^{-dn-n+3}.$$

The values of k_1–k_4 can be found with a minimum of effort, with the aid of Lenstra's 'Integer Programming Algorithm' (Lenstra 1983).

Assume that the values found in this manner result in Figure 6.4. In addition, the gradients of the 'teeth' are also given. Note that the slope of the teeth becomes steeper as i increases. The value of w_0/u_0 we are trying to determine will lie to the right of k_1/b_1, before the next zero of one of the n curves.

In order to focus our search effort, we will first divide this interval into subintervals. There are n lines/slope, which have a maximum of n^2 intersections within the interval. This will produce n^2 subintervals. Within these subintervals, the lines will not intersect and the sawtooth functions sb_i will be arranged in a certain order, depending on their respective magnitudes. This order also applies to the elements of a_i of the original knapsack vector. If these values constitute a super-increasing sequence, then within this interval we can search for a pair of numbers, w and u, whose ratio also lies within the interval. Once these are found, the knapsack algorithm has virtually been cracked, as it is now possible to reconstruct the original message X from S without any further problems.

Figure 6.4. Cracking the knapsack algorithm.

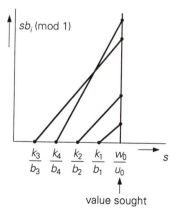

The knapsack algorithm can be made more complex by encrypting a super-increasing sequence to a vector B, using (v_1, u_1):

$$b_i = v_1 a_i \pmod{u_1} \text{ for all } i.$$

Vector B is enciphered for a second time with (v_2, u_2) into B':

$$b_i' = v_2 b_i \pmod{u_2} \text{ for all } i.$$

However, even this kind of system can be cracked.

Although confidence in the security of any system of this type has diminished considerably since Shamir's attack, alternative methods of improving its security are still being tested because it is such a simple system to use.

6.5 Public key systems based on elliptic curves

A common feature of the RSA system and the knapsack system is the fact that they are both based on so-called trapdoor functions. Calculating these functions is fairly straightforward. However, calculating the inverse function is far more difficult, unless additional (secret) information is available. A more recent development in the field of public key systems is based on the use of *elliptic curves*. The use of elliptic curves was first proposed by Miller (1986) and Koblitz (1987). We will now give an outline of how this method works, without delving too deeply into the underlying mathematics. For more details we refer the reader to Menezes (1993).

An *elliptic curve* given by F: $y^2 = x^3 + ax + b$ and defined for Z_p, where Z_p represents the set of integers between 0 and p, includes all pairs of numbers $(x, y) \in Z_p \times Z_p$ which satisfy:

$$y^2 = x^3 + ax + b \pmod{p}, \tag{6.35}$$

where p is a prime number, $p > 3$ and a and b are constants for which $4a^3 + 27b^2 \neq 0 \pmod{p}$.

Before explaining how elliptic curves are used for creating public key systems, it is necessary to consider the points lying on F and their relation to each other.

From eq. (6.35) the points on F can be found by determining $z = x^3 + ax + b \pmod{p}$ for every $x \in Z_p$, and then calculating the corresponding y. This will only be possible if z is a so-called *quadratic residue*, i.e. z must satisfy the equation $z = y^2 \pmod{p}$. *Euler's criterion*

(Kranakis 1986) states that z can only be a quadratic residue modulo p if and only if:

$$z^{(p-1)/2} = 1 \ (\text{mod } p).$$

By multiplying both sides of this equation by z, it follows that:

$$z^{(p+1)/2} = z \ (\text{mod } p).$$

We then find the roots of this equation are:

$$y = \pm \, z^{(p+1)/4}, \tag{6.36}$$

provided that p can be written as $p = 3 \ (\text{mod } 4)$. For this case we can now find the points on F.

Example

Consider the elliptic curve $y^2 = x^3 + x + 5$, which is defined over Z_{11} for $p = 11$. It can be verified that the constants $a = 1$ and $b = 5$ satisfy the condition stated for eq. (6.35). Since $p = 11 = 3 \ (\text{mod } 4)$, then the values for z (and consequently for y) are given for every x by eq. (6.36). These results are listed in Table 6.1. △

Clearly, we can find ten points for this case. In general, the total number of points will be of the order of the prime number p.

By selecting a suitable operator, the points on the elliptic curve F may be regarded as an *Abelian group*. This means that the operator + is chosen such that if $P \in F$ and $Q \in F$, then it holds that $P + Q \in F$.

Table 6.1. Points of the elliptic curve $y^2 = x^3 + x + 5$.

x	$z = x^3 + x + 5 \ (\text{mod } 11)$	quadratic residue	y	(x,y)
0	5	yes	4,7	(0,4) (0,7)
1	7	no		
2	4	yes	2,9	(2,2) (2,9)
3	2	no		
4	7	no		
5	3	yes	5,6	(5,5) (5,6)
6	7	no		
7	3	yes	5,6	(7,5) (7,6)
8	8	no		
9	6	no		
10	3	yes	5,6	(10,5) (10,6)

Assume $P = (x_1, y_1) \in F$ and $Q = (x_2, y_2) \in F$, then for the case $x_2 = x_1$ and $y_2 = -y_1$, we can select the operator such that $P + Q = O$; O is a point for which $P + O = P$, for all $P \in F$. This implies that the inverse of P must be $(x_1, -y_1)$. For all other cases, $P + Q = (x_3, y_3)$, with

$$x_3 = \sigma^2 - x_1 - x_2 \ (\text{mod } p), \tag{6.37}$$

$$y_3 = \sigma(x_1 - x_3) - y_1 \ (\text{mod } p)$$

and

$$\sigma = \begin{cases} (y_2 - y_1)/(x_2 - x_1), & \text{if } P \neq Q, \\ = (3x_1^2 + a)/(2y_1), & \text{if } P = Q. \end{cases} \tag{6.38}$$

All points of the elliptic curve can be generated with eqs. (6.37) and (6.38) by choosing an arbitrary point $P \in F$ as an initial starting point.

Example

Once again we will consider $y^2 = x^3 + x + 5 \ (\text{mod } 11)$. In the previous example we found that $P = (0,7) \in F$. We can now use eqs. (6.37) and (6.38) to calculate $2P, 3P, ..., 10P$. Let us start with the calculation of $2P$:

$$2P = (0,7) + (0,7).$$

Using eq. (6.38), we find that:

$$\sigma = (3 \times 0^2 + 1)/(2 \times 7) = 1/14 = 4 \ (\text{mod } 11),$$

since $4 \times 14 = 1 \ (\text{mod } 11)$. This result can then be used with eq. (6.37), yielding:

$$x_3 = 16 = 5 \ (\text{mod } 11),$$

$$y_3 = 4(0 - 5) - 7 = -27 = 6 \ (\text{mod } 11).$$

So we have now found the point (5,6), which is indeed a point on the elliptic curve.

The calculation of $3P$ is performed as follows:

$$3P = P + 2P = (0,7) + (5,6).$$

Using eq. (6.38) again, we now find that:

$$\sigma = (7 - 6)/(0 - 5) = -1/5 = 2 \text{ (mod 11)}.$$

Substituting this value in eq. (6.37) yields:

$$x_3 = 2^2 - 5 = -1 = 10 \text{ (mod 11)},$$

$$y_3 = 2(0 - 10) - 7 = -27 = 6 \text{ (mod 11)}.$$

This is the point (10,6).

In a similar manner the remaining points $3P$, $4P$, ... can be calculated. The results are given below:

$P = (0,7)$	$6P = (7,6)$
$2P = (5,6)$	$7P = (2,9)$
$3P = (10,6)$	$8P = (10,5)$
$4P = (2,2)$	$9P = (5,5)$
$5P = (7,5)$	$10P = (0,4)$

These points correspond exactly to those given in Table 6.1. △

The results found in the previous section can be used as a basis for the design of a cipher system. It is obvious that for a given P, the point αP can easily be calculated for any value of α. However, what does prove to be a difficult task is finding the value of α when P and αP are given. In the example given above, it is fairly straightforward to derive that $\alpha = 7$. The value of α can be found by generating the entire list P, $2P$, $3P$, ... etc. However, if the prime number p is very large, say of the order of 2^{160}, then the number points on the elliptic curve will be the same order of magnitude. It will therefore be totally impractical to generate the list P, $2P$, $3P$, ...

The El-Gamal cipher system based on elliptic curves

Let F be an elliptic curve defined for Z_p, with p a prime number and $p > 3$; P is a point on the elliptic curve: $P \in F$. The number α is a secret exponent. Q is defined as: $Q = \alpha P$ ($Q \in F$). The public key consists of P and Q. The value of α is the secret key.

Let the message to be enciphered be denoted by M, for which $M = (u_1, u_2)$ and $M \in F$. If k is a random number, then the encipherment is defined as:

$$C = e(M,k) = (v_1, v_2), \tag{6.39}$$

with

$$v_1 = kP \text{ and } v_2 = M + kQ.$$ (6.40)

Decipherment can be performed according to:

$$d(C,\alpha) = v_2 - \alpha v_1.$$ (6.41)

Since $\alpha P = Q$, it is obvious that $d(C,\alpha)$ will result in the original message M. The strength of the algorithm relies on the fact that it is absolutely impractical to find α, when only P and Q are known.

Example

Assume $P = (0,7)$ and $\alpha = 3$. Then it follows that $Q = 3(0,7) = (10,6)$.

Suppose the message to be enciphered is given by $M = (2,9)$. The random number k is given by $k = 6$. The cipher text will now consist of two elements:

$$v_1 = 6P = (7,6),$$

$$v_2 = M + kQ = (2,9) + 6(10,6) = (10,6).$$

For decipherment, the values of v_1, v_2 and α must be known:

$$d(C,3) = v_2 - 3v_1 = (10,6) - 3(7,6) = (10,6) + 3(7,5) = (2,9).$$

Here we employed the fact that $-(7,6) = (7,-6) = (7,5)$.

A disadvantage of this algorithm is that the messages to be enciphered must be points on the curve F. In the Menezes–Vanstone cipher system, the messages may be any random pair of numbers.

The Menezes–Vanstone cipher system based on elliptic curves

Let F be an elliptic curve defined on Z_p (p is a prime number, $p > 3$). P is a point on the elliptic curve: $P \in F$. The number α is a secret exponent. Q is defined as $Q = \alpha P$. P and Q are made public.

Let the message which is to be enciphered be denoted as $M = (u_1,u_2)$. If k is a random number, then the encipherment is given by

$$C = e(M,k) = (y_0,y_1,y_2),$$ (6.42)

with

$$y_0 = kP, (c_1,c_2) = kQ, y_1 = c_1 u_1 \pmod p, y_2 = c_2 u_2 \pmod p.$$ (6.43)

Decipherment is performed by calculating

$$d(C,\alpha) = (y_1 c_1^{-1} \pmod{p}, y_2 c_2^{-1} \pmod{p}), \tag{6.44}$$

in which

$$\alpha y_0(c_1,c_2). \tag{6.45}$$

○

Example

Assume the same values for P, Q, k and α as in the previous example. However, now the message to be enciphered is $M = (3,4)$, which is not a point on the elliptic curve F. The elements of the cipher text are subsequently found to be:

$$y_0 = kP = 6(0,7) = (7,6),$$

$$(c_1,c_2) = kQ = 6(10,6) = (2,9),$$

$$y_1 = c_1 u_1 = 2 \times 3 \pmod{11} = 6,$$

$$y_2 = c_2 u_2 = 9 \times 4 \pmod{11} = 3.$$

From this it follows that:

$$C = ((7,6),6,3).$$

The deciphering is performed as follows:

$$(c_1,c_2) = \alpha y_0 = 3(7,6) = (2,9),$$

$$M = (y_1 c_1^{-1}, y_2 c_2^{-1}) = (6/2, 3/9) = (3,4). \qquad \triangle$$

A disadvantage of both of these methods is the fact that the ciphertext becomes twice as long as the plain text. On the other hand, however, in these methods smaller prime numbers can generally be used than in the RSA system. Furthermore, the size of the public key is limited and, consequently, the complexity of the calculations is also limited. We can state that, in general, encipherment and decipherment based on elliptic curves requires much less time than that based on the RSA.

7

Authentication and integrity

7.1 Protocols

Often, the availability of a secure cryptographic algorithm for enciphering data, as described in the previous sections, is not sufficient in itself. In addition, we must also agree on exactly how the information is to be exchanged. These agreements are called the *cryptographic protocol* and offer a set of rules for exchanging information, which prevent fraud by either of the communicating parties or an intruder. Many eventualities, which we would rather avoid when exchanging information, are conceivable.

Suppose A (Alice) and B (Bob) wish to exchange information. Although they use a safe cryptographic algorithm (for instance, a public key system), they still have no guarantee of protection against:

– tapping/listening in by C (Charles);
– distortion of the message by Charles;
– the possibility of Charles posing as Bob;
– the possibility of Bob sending a message to Alice, but denying it later;
– the possibility of Alice denying the fact that she received a message, even if she has;
– the possibility that Alice may invent a message she really didn't receive etc..

Obviously, some kind of protection against occurrences is desirable. In practice it is almost impossible to design a protocol which can prevent any form of misuse, as we will now demonstrate with the following example.

Suppose Alice wishes to send a message M to Bob, with some means of verifying that Bob has received the message correctly, without any kind of distortion. The system must then be secured against intruders. This can be accomplished with, for instance, the RSA system. Let P_A and S_A denote the public and secret keys of Alice and P_B and S_B the public and secret keys of Bob, respectively. In this section we will use the notation $eP_B(M)$ etc. for the result obtained after encipherment of message M with key P_B. Consider the following protocol.

Protocol I (see Figure 7.1)

(1) Alice encrypts the message M with Bob's public key P_B.
   She then sends this to Bob, along with her own address (otherwise Bob does not know who sent the message): $C1 = eP_B(M)//A$.
(2) Bob will decipher $C1$ with his secret key S_B, thus retrieving M and A. Then, M is enciphered again, but now using the public key P_A, and this is sent together with Bob's address back to Alice: $C2 = eP_A(M)//B$.
(3) Now Alice can decrypt $C2$ and check whether Bob received the message M without corruption.

The first impression is that this protocol is safe. However, a closer examination reveals that an intruder, Charles, can still obtain the message M. He can accomplish this with the following method (see Figure 7.2):

(1) Charles first intercepts $C1$ and replaces address A, which is not enciphered, with his own address C. Bob will now receive: $C1' = eP_B(M)//C$.
(2) Bob deciphers M and C and returns $C2' = e\,P_C(M)//B$ to Charles, according to the agreed protocol.
(3) Finally, Charles deciphers $C2'$ with his own secret key S_C, thus obtaining M. In order to let Alice believe that everything has gone according to plan, he can send $C2 = eP_A(M)//B$ to her.

The reason that this protocol falls short in this way is the fact that the addresses of the transmitters are not enciphered. Thus, an intruder can easily

Figure 7.1. Protocol I.

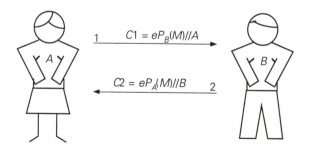

substitute any address with his own. We can therefore improve this protocol by using two encryptions, which prevent the addresses appearing in the plaintext (see Figure 7.3).

Protocol II

(1) Alice sends $C1 = eP_B[eP_B(M)//A]$ to Bob.
(2) Bob deciphers this with S_B and obtains M and A. He then returns $C2 = eP_A[eP_A(M)//B]$ to Alice.
(3) Alice deciphers $C2$ in order to verify that Bob has received the message

Figure 7.2. An attack on protocol I.

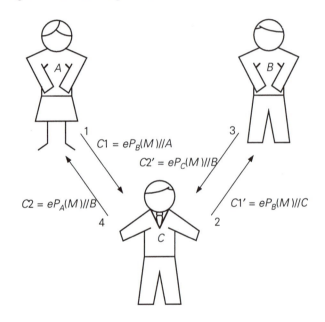

$C1 = eP_B(M)//A$

$C2' = eP_C(M)//B$

$C2 = eP_A(M)//B$

$C1' = eP_B(M)//C$

Figure 7.3. Protocol II

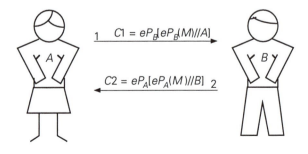

$C1 = eP_B[eP_B(M)//A]$

$C2 = eP_A[eP_A(M)//B]$

correctly.

Despite the fact that here, the addresses have been enciphered, Charles can still outsmart Bob and Alice, as is demonstrated by the following steps (see also Figure 7.4):

(1) Charles intercepts $C1$, adds his own address and enciphers the entire text with Bob's public key: $eP_B(C1//C) = eP_B[eP_B[eP_B(M)//A]//C]$.
(2) After deciphering this, Bob finds $[eP_B(M)//A]$ and C. According to protocol, he will now return to Charles: $eP_C[eP_C[eP_B(M)//A]//B]$.
(3) From this, Charles can derive $eP_B(M)$ and send $eP_B[eP_B(M)//C]$ back to Bob.
(4) Bob will now find M and C and send $eP_C[eP_C(M)//B]$ to Charles, all according to protocol.
(5) Charles can decipher this with his secret key S_C and thus obtain M. Alice can be deceived by sending $eP_A[eP_A(M)//B]$ to her.

In practice, it is not an easy task to determine whether a given protocol is really secure and that there is absolutely no possibility of an intruder obtaining the original message, especially where complex protocols are concerned. The last few years have seen the development of techniques based on artificial intelligence for the evaluation of this type of protocol as part of the designing process.

Figure 7.4. An attack on protocol II.

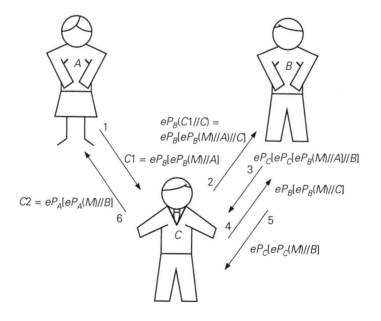

We will now provide a protocol which is demonstrably secure.

Protocol III (see Figure 7.5)

(1) Alice sends $C1 = eP_B(M//A)$ to Bob.
(2) From this, Bob can determine M and A return $C2 = eP_A(M//B)$ to Alice.
(3) Alice may now verify the transmission of the message.
This protocol can be considered as a special form of authentication. A can verify whether B has received the message without any form of distortion.

In general, we can distinguish between message integrity, entity authentication and message authentication.

❑ *Message integrity*
Message integrity is concerned with preventing the data being manipulated. It involves not only methods of detecting whether a stored or transmitted message has been altered, but also whether the message has been replayed (tapped) by an intruder and how this could be prevented.

❑ *Entity authentication*
Literally this means determining the true identity of a person, system, etc. Here, we are interested in the identity of the transmitter of a message. If B receives a message from A, he would like some guarantee that this message was actually sent by A and not by a third party C, posing as A. Similarly, A would like some means of ensuring that a message sent to B is actually received by B.

❑ *Message authentication*
This may be regarded as a combination of message integrity and entity authentication. Both parties are able to verify each other's authenticity and whether the data are still undamaged.

Figure 7.5. Protocol III.

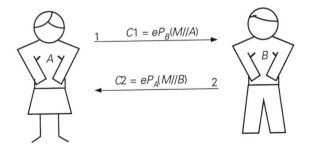

The last example is, in fact, an example of message authentication, though in a rather basic form. Alice can verify whether Bob has received the message without any kind of corruption. In addition, she will also have some guarantee that it was actually Bob who received the message, since she used Bob's public key for encipherment and only Bob can decipher the message in order to send it back to Alice.

The authentication remains unilateral, though, as Bob has no means of verifying whether he is actually dealing with Alice; anyone can claim to be Alice. Therefore, a better solution would be to use so-called digital autographs, which we will discuss later.

7.2 Message integrity with the aid of Hash functions

In Section 7.1 we encountered a form of message integrity, which used the entire message for verification. However, there are faster methods of checking message integrity which are based on a single figure, derived from the message. This figure can be calculated with one-way functions. The message (or a part of the message) serves as the input of the one-way function. The result of the function can be used for verification and should be transmitted separately, since anyone can calculate this value. Often, *Hash functions* are used for this purpose.

We came upon the concept of one-way functions in chapter 6. A one-way function will irreversibly map a set of input values onto a set of output values. This means that the output values are easily computed, but it is virtually impossible to compute the input values, based on given output values alone.

A *Hash function* is a special function in the sense that it maps a set input values of variable length onto a set of output values of a constant length. The Hash function is publicised so anyone can calculate the Hash result and verify a received message.

A wishes to send a message M to B (see Figure 7.6). To guarantee the integrity of the message, A calculates the Hash result of message M, i.e. $h(M)$. After B has received the message, he will also determine the Hash result of the received message and compare this to the first result $h(M)$, which is sent to him separately. In this way, he can verify that the message has been received correctly. If the data of the message had been manipulated in some way, then B would receive M' instead of M and the Hash result $h(M')$ would differ from $h(M)$.

The principal characteristic of a Hash function lies in the requirement that it must be practically impossible to create a message M' whose corresponding Hash code $h(M')$ is equal to that of a given message M: $h(M')$ $\neq h(M)$. Suppose the Hash code is represented by n bits. Then there are 2^n possible Hash codes. Therefore, the chance that a cryptanalyst can find an M' which produces the same Hash code as M is 2^{-n}. If he makes k attempts at finding such a text, this chance will increase to $1 - (1 - 2^{-n})^k$, which is approximately equal to $k2^{-n}$. Suppose a chance of success no larger than 10^{-6} after 2^{55} attempts is acceptable (this corresponds to an exhaustive key search for the DES). Then the total number of Hash codes must be at least 2^{35} (i.e. 35 bits).

A Hash function must also be *collision free*. This means that it must be impossible to construct two different messages with the same Hash code (this differs with the previously discussed case in that here M is not given beforehand). The search for pairs of messages, M and M', for which $h(M') = h(M)$, is called a *birthday attack*. The underlying principle of a birthday attack originates from the so-called birthday paradox.

Assume we have a group of 23 people. The probability that two or more persons have the same birthday is slightly larger than 50%. This is a lot higher than may at first be expected. In order to answer the question why this is so, we must first consider this problem in a general form. Assume we have r random numbers (here $r = 23$ persons), which range from 1 to n (in this example $n = 365$, the number of days in a year). What is the probability of two or more of the r numbers having the same value? This can be computed in the following manner. First, we note all of the n values. We then repeatedly take one of the r numbers, consider its value and cross this value off the list of n values. If a value has already been crossed off, then obviously we have found two equal numbers and the procedure can be terminated. Consider the probability that this *does not* occur. The probability that the first number selected from the group of r numbers is not equal to a previously selected number must be 1, since there is no previously selected

Figure 7.6. Message integrity based on a Hash function.

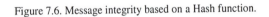

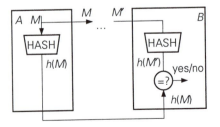

number. When the second number is selected, there will still be $n - 1$ values on the list. Therefore, the probability that the second number is equal to one of the remaining values of the list is equal to $(n - 1)/n$. The probability of the third number corresponding to one of the remaining values is $(n - 2)/n$, etc. Thus, the probability of finding no two equal numbers is:

$$p(\text{no equality}) = \frac{n}{n} \cdot \frac{n-1}{n} \cdot \frac{n-2}{n} \cdot \ldots \cdot \frac{n-r+1}{n} = \frac{n!}{(n-r)!n^r}. \qquad (7.1)$$

With $n = 365$ and $r = 23$, this probability is equal to 0.493. Therefore, the chance that two or more numbers are the same is a little more than 50%. In general, the probability of success is approximately equal to:

$$p(\text{success}) \approx 1 - e^{-r^2/2n} \quad (r \ll n). \qquad (7.2)$$

This means that if the probability of finding two messages with the same Hash code may not exceed 10^{-6} after 2^{55} attempts, then $n > 2^{128}$ (i.e. the Hash code length must be at least 128 bits).

It is important that the Hash code is transported from A to B safely. This may be ensured physically, for instance by a courier, or automatically. In the latter case, the Hash code must be transmitted in an enciphered form, for example with a symmetrical DES algorithm. An intruder is not able to manipulate the Hash code in a purposeful manner, since he cannot predict the result after decipherment.

A good Hash function can also resist a known-plaintext-attack.

DES Hash

A feasible method for creating a Hash function is by using the DES as a one-way function, as in Figure 7.7. M is enciphered with the DES and a secret key. This result is then added to the original M, modulo 2. It is evident that C can be generated from M without much effort, but that the reverse operation, i.e. finding M from C' is almost impossible, even if K is given. The DES can be utilised in the following manner to generate a Hash function.

Figure 7.7. DES as a one-way function.

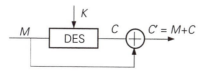

Consider Figure 7.8. The total message *M* is divided into blocks of 64 bits. If the length of the message is not a multiple of 64 bits (a requirement imposed by the DES algorithm), the message can be padded with extra zeros, i.e. extra zeros are concatenated with the message until the length is a multiple of 64. This results in a series of blocks, denoted by *M*1, *M*2, *M*3, etc., which are used as the input for the DES algorithm in a one-way implementation. Each block is encrypted with a key, which is based on the result of the previous computations. It holds for every *i* that:

$$Ki = U(Ci), \tag{7.3}$$

$$Ci = EK(i-1)(Mi) + Mi. \tag{7.4}$$

$U(.)$ is an arbitrary function which transforms the output of each one-way function to a key. An initial key *K*0 is used for the first block *M*1. The one-way function is applied repeatedly until all the blocks of data have been enciphered. The Hash result is found by taking the *l* left-hand bits of the result *Kn*, where *l* still remains to be defined.

MD5 Message Digest

Another example of a Hash function is given by the MD5 Message Digest algorithm, developed by Rivest. The algorithm will compress a message with an arbitrary length to a 128-bit Hash result (*fingerprint of message digest*) in such a way that it is impossible to find two messages which produce the same result or to generate a message which will produce a pre-defined Hash result.

In contrast to the above example, where the DES Hash uses blocks of 64 bits, input blocks of 32 bits are now used. Let us consider the algorithm in detail.

(1) Assume we have a message *M* of *k* bits. The length of the message is extended by padding it with zeros, until it is 64 bits short of a multiple

Figure 7.8. The DES Hash.

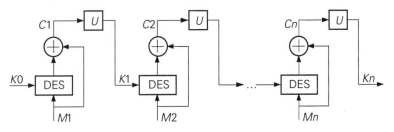

of 512 bits. These last 64 bits are filled with a binary representation of k and the total length becomes a multiple of 512. The message is divided into blocks of 32 bits, which are noted as 32-bit words $M1$, $M2$, $M3$, etc., for subsequent operations. Note that now the message is a multiple of 16 words (of 32 bits each). See Figure 7.9.

(2) Four logical functions are defined, with which 32-bit output blocks are generated from 32 input bits. These functions are given as:

$$f(X,Y,Z) = (X \wedge Y) \vee ((\neg X) \wedge Z),$$

$$g(X,Y,Z) = (X \wedge Y) \vee (Y \wedge (\neg Z)),$$

$$h(X,Y,Z) = X + Y + Z,$$

$$i(X,Y,Z) = Y + (X \wedge (\neg Z)),$$

where X, Y and Z are 32-bit words. The function $h(X,Y,Z)$ is an XOR function. The function $g(X,Y,Z)$ can be described as a kind of 'majority function': if a bit at a certain position of at least two of the words X, Y or Z is equal to 1, then the function will also have a 1 at the corresponding position. The functions $f(X,Y,Z)$ and $i(X,Y,Z)$ can be interpreted in the same manner.

(3) Now the Hash result is computed by processing the input blocks in four rounds, each comprising 16 steps.
  – Round 1
  Let $[A\ B\ C\ D\ i\ s\ t]$ represent the operation $A = B + \{[A + f(B,C,D) + Mi + t] \lll s\}$, where t is a constant and where $\lll$ s stands for a cyclic shift of a 32-bit word over s positions.
  The initial values of A, B, C and D are equal to the hexadecimal values:

  $A = 01\ \ 23\ \ 45\ \ 67,$

Figure 7.9. Initial processing for the MD5 Message Digest algorithm.

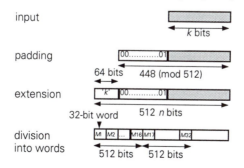

$B = $ 89 AB CD EF,

$C = $ FE DC BA 98,

$D = $ 76 54 32 10.

Table 7.1(a) demonstrates the 16 operations which are performed on the data.

Table 7.1. Operations of the MD5 Message Digest algorithm.

(a)	(b)
[$A B C D$ 0 7 D76AA478]	[$A B C D$ 1 5 F61E2562]
[$D A B C$ 1 12 E8C7B756]	[$D A B C$ 6 9 C040B340]
[$C D A B$ 2 17 242070DB]	[$C D A B$ 11 14 265E5A51]
[$B C D A$ 3 22 C1BDCEEE]	[$B C D A$ 0 20 E9B6C7AA]
[$A B C D$ 4 7 F57C0FAF]	[$A B C D$ 5 5 D62F105D]
[$D A B C$ 5 12 4787C62A]	[$D A B C$ 10 9 02441453]
[$C D A B$ 6 17 A8304613]	[$C D A B$ 15 14 D8A1E681]
[$B C D A$ 7 22 FD469501]	[$B C D A$ 4 20 E7D3FBC8]
[$A B C D$ 8 7 698098D8]	[$A B C D$ 9 5 21E1CDE6]
[$D A B C$ 9 12 8B44F7AF]	[$D A B C$ 14 9 C33707D6]
[$C D A B$ 10 17 FFFF5BB1]	[$C D A B$ 3 14 F4D50D87]
[$B C D A$ 11 22 895 CD7BE]	[$B C D A$ 8 20 455A14ED]
[$A B C D$ 12 7 6B901122]	[$A B C D$ 13 5 A9E3E905]
[$D A B C$ 13 12 FD987193]	[$D A B C$ 2 9 FCEFA3F8]
[$C D A B$ 14 17 A679438E]	[$C D A B$ 7 14 676F02D9]
[$B C D A$ 15 22 49B40821]	[$B C D A$ 12 20 8D2A4C8A]
(c)	(d)
[$A B C D$ 5 4 FFFA3942]	[$A B C D$ 0 6 F42922244]
[$D A B C$ 8 11 8771F681]	[$D A B C$ 7 10 432AFF97]
[$C D A B$ 11 16 6D9D6122]	[$C D A B$ 14 15 AB9423A7]
[$B C D A$ 14 23 FDE5380C]	[$B C D A$ 5 21 FC93A039]
[$A B C D$ 1 4 A4BEEA44]	[$A B C D$ 12 6 655B59C3]
[$D A B C$ 4 11 4BDECFA9]	[$D A B C$ 3 10 8F0CCC92]
[$C D A B$ 7 16 XF6BB4B60]	[$C D A B$ 10 15 FFEFF47D]
[$B C D A$ 10 23 BEBFBC70]	[$B C D A$ 1 21 85845DD1]
[$A B C D$ 13 4 289B7EE6]	[$A B C D$ 8 6 6FA87E4F]
[$D A B C$ 0 11 EAA127FA]	[$D A B C$ 15 10 FE2CE6E0]
[$C D A B$ 3 16 D4EF3085]	[$C D A B$ 6 15 A3014313]
[$B C D A$ 6 23 04881D05]	[$B C D A$ 13 21 4E0811A1]
[$A B C D$ 9 4 D9D4D039]	[$A B C D$ 4 6 F7537E82]
[$D A B C$ 12 11 E6DB99E5]	[$D A B C$ 11 10 BD3AF235]
[$C D A B$ 15 16 1FA27CF8]	[$C D A B$ 2 15 2AD7D2BB]
[$B C D A$ 2 23 C4AC5665]	[$B C D A$ 9 21 EB86D391]

When these operations are completed, round 2 commences.

– Round 2

This round employs the logical function $g(X,Y,Z)$. Here, 16 operations are performed, according to

$$[A\ B\ C\ D\ i\ s\ t]: A = B + \{(A + g(B,C,D) + Mi + t) \lll s\}.$$

See Table 7.1(*b*).

– Rounds 3 and 4

Rounds 3 and 4 are based on the functions $h(X,Y,Z)$ and $i(X,Y,Z)$, respectively, where the operations are described by

$$[A\ B\ C\ D\ i\ s\ t]: A = B + \{(A + h(B,C,D) + Mi + t) \lll s\},$$

and

$$[A\ B\ C\ D\ i\ s\ t]: A = B + \{(A + i(B,C,D) + Mi + t) \lll s\}.$$

The results of the 16 operations are defined as in Table 7.1(*c*) and (*d*).

(4) Final result: The resulting values of A, B, C and D at the end of round 3 are used as initial values for processing the next 16 words of 32 bits, according the steps described above. The process is repeated until all 32-bit words have been used. The Message Digest of 128 bits is formed by the four values of A, B, C and D which are finally obtained after these operations.

Although the implementation of the algorithm may seem fairly straightforward, from a functional point of view it is extremely complex in order to meet the necessary requirements. The number of operations which are required to generate two messages which produce the same Message Digest is estimated at 2^{64}. The complexity of finding a message with a pre-defined Message Digest is even larger, of the order of 2^{128}. We can therefore consider finding two messages with the same Message Digest and finding a message with a pre-defined Message Digest to be impossible tasks.

7.3 Entity authentication with symmetrical algorithm

In the previous section we discussed methods of verifying message integrity. However, when determining the source of a message, i.e. entity authentication, two groups of methods are used: those based on symmetrical and asymmetrical algorithms. Here, we will consider methods which rely on symmetrical algorithms.

For entity authentication with symmetrical algorithms, an inquirer will request a second party to perform some kind of cryptographic operation involving a secret key on unpredictable data sent by the inquirer. The inquirer can compare this result with that result obtained by performing the same cryptographic operation himself to establish whether the entity is authentic or not. If the two results correspond well, then the second party must indeed have access to the secret key, which belongs specifically to the second party.

Consider Figure 7.10 (two-way challenge–response), where A is verifying B's authenticity. A sends an unpredictable value R_A (the challenge) to B; R_A is, for instance, a random number. B will then generate a value $oK(R_A)$ (the response) with a symmetrical one-way function and a secret key K. In the mean time, A also generates this value. For an example of this situation in practice, A can be thought of as a central computer, which may only be accessed by authorised users.

In the above example, A only verifies the authenticity of B and not vice versa. The situation in which a mutual test of each other's authenticity is performed is depicted in Figure 7.11 (three-way challenge–response). In Figure 7.11, A can establish the authenticity of B by considering the result $oK(R_A)$, which B has generated from R_A. As B will return not only the result

Figure 7.10. Two-way challenge–response.

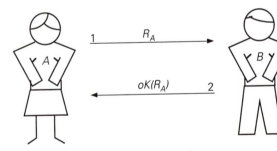

Figure 7.11. Three-way challenge–response I.

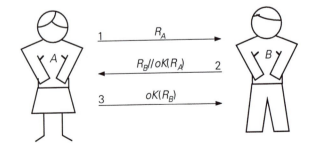

$oK(R_A)$, but in addition another random value, R_B, B is now able to verify A's authenticity with the value $oK(R_B)$ generated and returned by A.

A weakness of this arrangement is its sensitivity to a so-called *reflection attack*. If an intruder C generates two parallel sessions, he can mislead A into believing he is dealing with B, when, in fact, he is dealing with C. Suppose A sends the challenge R_A. However, this time it is not B, but intruder C who receives the challenge. C immediately returns the same R_A to A, as if it were sent by B. Then A transmits $oK(R_A)$ and a new challenge $R_{A'}$. Again C responds immediately, now with the same $oK(R_A)$. This is the correct response for A and therefore A wrongly assumes he is dealing with B. Obviously, in this case, entity authenticity cannot be guaranteed. This problem can be solved by sending the names of A and B along with the authentication messages. Consider Figure 7.12. Here, the result of the one-way functions also depends on the relevant address and the system is secure against reflection attacks.

In all of the above examples, we assumed that before the authentication process was initiated, A and B already had the secret key K. However, this is not always the case. Figure 7.13 offers a solution for those cases in which A and B do not have access to a secret key prior to the authentication process. Now a third party, or so-called *trusted party TP*, is introduced, who can communicate with A and B using two secret keys K_1 and K_2. The authentication process will now proceed according to the following steps.
(1) A sends a random number R_A to B.
(2) B sends A's name and a reference number N_B to C, the 'trusted party'. This number will prevent C from returning a simple replay to B. Thus, B sends $A//N_B$ to C.
(3) C will now construct a complicated message using a symmetrical cipher algorithm and return this to B: $EK_2[EK_1(B//F)//N_B//A//F]$. In this message, F contains the secret key K, which is used for the actual authentication. Note that A cannot decipher what C sends to B, since the

Figure 7.12. Three-way challenge-response II.

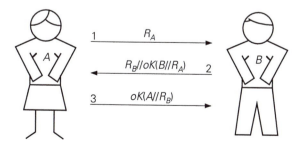

key K_2 is only known to B and C and not to A. In turn, B can retrieve F and (of course) B from the above, but will find it impossible to derive the key K_1, which is shared only by A and C, from $EK_1(B//F)$. B can, however, find the authentication key K.

(4) B will now also generate a random number, R_B and subsequently send to A: $EK_1(B//F)$, which had previously been received from C. In addition, B also sends $EK(R_A//R_B)$ to A. Thanks to C, B now knows K and is therefore able to generate this last message.

Once A has received all this, he will use the secret key K_1 (known only to A and C) to obtain the authentication key K from F, by deciphering $B//F$. Now A has discovered K, he can decipher $EK(R_A//R_B)$, which will finally produce the values of R_A and R_B.

Note that if an intruder D were to assume B's place after step 1, A would receive $EK_1(C//F)$, instead of $EK_1(B//F)$. Thus A would then realise that he was communicating with someone other than B.

(5) After A has decrypted R_B, he sends this back to B, who can then conclude that no-one other than A could have sent this message, since the key K, with which A found the random number R_B, was enciphered by C, with a secret key known exclusively to A and C.

Figure 7.13. Authentication involving a 'trusted party'.

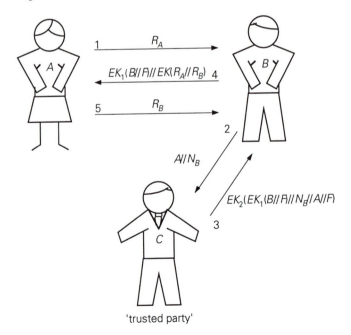

7.4 Message authentication with a message authentication code (MAC)

Message authentication can be regarded as a combination of message integrity and entity authentication. Message authentication can be realised by employing a secret function, instead of a public function (as is the case for the determination of the Hash code). This method is followed for the so-called MAC (message authentication code), where a symmetrical algorithm is used as a one-way function. Unlike the case of the Hash functions, the result can be transmitted with the message and does not need to be sent separately.

In Figure 7.14, $mgK(M)$ represents the MAC generated by the symmetrical algorithm and the key K, for given message M. Now B can check the integrity of the message by generating a MAC in the same manner as A and comparing this with the received value. If the values are identical, then B can conclude that the message is intact and that it was indeed sent by A, since the secret key is only available to A (this statement implies correct entity authentication). It is evident that the authenticity will depend on the cryptographic strength of the algorithm and also on the level of confidentiality with respect to the secrecy of the key K. This key must therefore be made available to both parties in a secure manner in advance.

A good MAC should be able to withstand a chosen-plaintext attack and also the previously mentioned birthday attack.

The use of a MAC will only prove beneficial under the condition that both parties trust each other entirely. If this were not the case, then A might send a message M to B, after which B declares receiving a message M', which he generated himself, including the corresponding MAC.

A MAC can be generated by employing the DES algorithm in CBC mode (see Figure 4.13). The key K must remain confidential. The initial vector, however, can be made public. The value of the MAC is equal to the 32 left-hand bits of the last block of the ciphertext. The MAC will therefore depend

Figure 7.14. Message authentication with MAC.

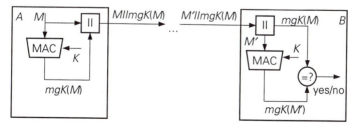

on both the message and the secret key K.

It is obvious that the probability of an intruder generating a message M' with the same MAC as that of M, is 2^{-32}. With respect to the discovery of the key with which the MAC is generated, an exhaustive key search will take the same number of operations and computing power as previously discussed for the DES algorithm.

A triple-DES is often used for the last block of data, in which case, an exhaustive key search would be impossible.

7.5 Message authentication with digital signatures

In the previous section, message authentication between A and B relied entirely on their mutual trust in each other. Provided that A and B have such a trusting relationship, this means of authentication provides adequate protection against intrusion by a third party.

Specific problems arise when the relationship between A and B cannot be based on confidentiality. B can pretend to have received a message from A, which was, in fact, never sent. For example, A sends a bank cheque to B via a digital communication network. Subsequently, B increases the amount of the bank draft and then claims that this (higher) sum was sent by A. In the same manner, B can also modify contracts and other important documents, which are transmitted via communication networks, to his own advantage.

The methods discussed so far provide protection against a third party, but in no way do they eliminate this kind of manipulation by the receiver B. This also applies to manipulation by A. For instance, A might attempt to evade the obligations of a contract by maintaining that he did not send the contract. In these situations, an impartial person cannot judge who is right; A claims that he did not send the message B holds and that B generated it himself, while B maintains that A did send the message but is trying to avoid his obligations.

In this kind of situation, a *digital signature* is required, which is generated by A and can only be verified and not generated by B. A means of accomplishing this is provided by the RSA algorithm.

The RSA algorithm was discussed in the previous chapter. The encipher and decipher operations are represented by:

$$C = M^e \pmod{n}$$

and

$$M = C^d \pmod{n},$$

respectively. The values of (e,n) are public, whereas d is only known to an authorised receiver. The algorithm is based on the principle that an unlawful receiver is not able to derive the value of d from e and n. Therefore, only the authorised receiver can decrypt the ciphertext.

In general, we can state that for public key systems

$$C = eP_A(M) \text{ and } M = dS_A(C),$$

where P_A and S_A represent the public and secret keys of A. All messages addressed to A are enciphered with P_A and A can decipher any received message with S_A. See Figure 7.15(a).

If, contrary to the above, S_A is used for encipherment and P_A is used for decipherment, the same diagram can be used to obtain digital signatures. Consider Figure 7.15(b). Now anyone can decipher data, but only the authorised transmitter can encipher a message. Only A has access to the secret key S_A and it is impossible to derive S_A from the public key P_A. The ciphertext C can serve as a guarantee that the text was sent by an authorised person for the receiver. After all, no-one else could have generated C, since only an authorised source has access to S_A. If another S_A were used instead, the deciphered message would not make sense. Thus, it is possible to authenticate a source with an asymmetrical algorithm, which was not possible with the methods described in the previous section. This method

Figure 7.15. The use of public key systems for (a) data protection and (b) implementing digital signatures.

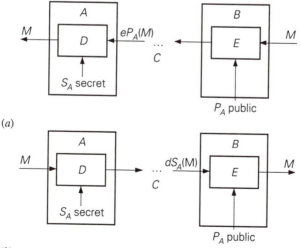

(a)

(b)

described can also be used for verification of message integrity.

A good example of this is the authentication of software. By supplying software with digital signatures, users are able to verify that no unauthorised alterations have been made to the software. Obviously, the public key must be made available to the software users. The same procedure can also be used for protecting software updates against imitations.

Returning to Figure 7.15(a), we must note that still no security measures have been taken with respect to the actual message itself, since anyone is able to decipher the ciphertext.

It is possible to use a public key system for the security of both the data and digital signatures, provided that encipherment followed by decipherment results in exactly the same text as decipherment followed by encipherment. This is the case for the RSA algorithm, as shown in Figure 7.16.

Both A and B have their own individual secret and public keys. Before A sends a message, a digital signature is added, which is encrypted with the secret key S_A. This result is then enciphered with the public key P_B owned by the receiver. Once the message has been transmitted, it can be deciphered with the secret key S_B. The data exchange is finally completed by authentication with the public key P_A.

Figure 7.16. Encipherment of messages with digital signatures based on a public key system.

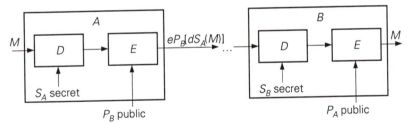

Figure 7.17. Message authentication with digital signatures.

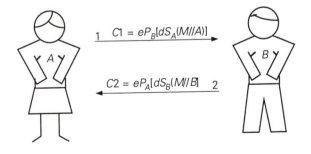

Let us return to protocol III of Section 7.1. The protocol as described by Figure 7.5 can be improved by adapting it to the protocol of Figure 7.17. Unlike the situation in Figure 7.5, mutual authentication is now possible.

To obtain a digital signature the data is often pre-processed, for instance by applying a Hash function. This can increase the efficiency. If the authentication is based on the RSA algorithm, this is especially desirable, since the arithmetic complexity of this algorithm requires that the data remain as compact as possible.

In Figure 7.18, the message M is first fed through a Hash function, resulting in $h(M)$. This result is then enciphered into $eS_A[h(M)]$, which is noted in an abbreviated form as $sS_A(M)$, which is then sent to B. Then B can decipher $sS_A(M)$ with P_A and retrieve $h(M)$. The value of $h(M)$ is finally compared to B's own calculations of the Hash result and provided that they correspond, B can conclude that (1) the message M is intact and (2) it originated from an authentic source, since only A knows the secret key with which the Hash result can be enciphered correctly in to $sS_A(M)$. Obviously, the actual message M may also be enciphered. This will lead to the diagram shown as Figure 7.19.

Figure 7.18. Digital signature of a Hash result.

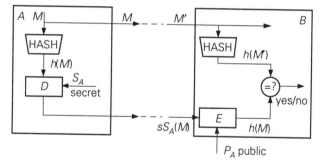

Figure 7.19. Digital signature combined with encryption.

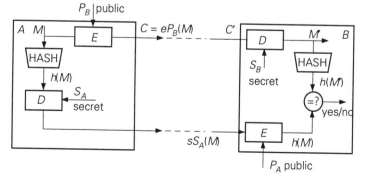

In a situation where digital signatures are used as described above (i.e. the so-called text hashing mode of signature), obviously the message itself must also be transmitted in addition to the signature. There are several methods which do not require M to be sent separately (text recovery mode of signature), which we will not discuss here, however. In these cases, the digital signature not only confirms the authenticity of the message and source, but, in addition, the message itself can be deduced from the digital signature.

Digital Signature Algorithm (DSA)

The RSA is an extremely effective algorithm for creating digital signatures. However, it has the disadvantage that it can also be used for encryption. Governments are becoming more and more concerned with the increasing use of cryptography, since it frustrates their efforts for monitoring organised crime. See Section 8.5. Anyone who uses the RSA for generating digital signatures can also use the algorithm for encryption. For this reason, a lot of effort is being spent on the development of algorithms which can be used for generating Hash codes and digital signatures, but which cannot be used for enciphering and deciphering data.

Such an algorithm was proposed by the NIST (National Institute of Standards and Technology in the USA) in 1991 under the name Digital Signature Algorithm (DSA) and has become a standard under the name Digital Signature Standard (DSS).

DSA

The algorithm starts with the following calculations. First, two prime numbers p and q are generated. For p it holds that $2^{i-1} < p < 2^i$, in which i is a multiple of 64 and $512 \leq i \leq 1024$. Furthermore, q is divisor of $p - 1$ and it must hold that $2^{159} < q < 2^{160}$. Then a random number h, with $h < p - 1$, is chosen such that:

$$g = h^{(p-1)/q} > 1.$$

The value of x is such that $x < q$ and for y:

$$y = g^x \pmod{p}. \tag{7.5}$$

The algorithm also uses a Hash function $h(M)$. The DSS is based on the Secure Hash Algorithm (SHA), although, at this point, the exact type of function is not of great importance.

The numbers p, q and g are made public and can, in fact, be used by a large number of users at once. Each user has his own secret key x. The public key is y. Both x and y can be used more than once. However, the value of the parameter k must be generated freshly for each new signature.

Suppose Alice wishes to place a digital signature over her message M. She must proceed as follows. First, Alice generates a secret random number k, $k < q$, and calculates

$$r = g^k \bmod p \;(\bmod\; q), \tag{7.6}$$

$$s = k^{-1}(h(M) + xr) \;(\bmod\; q). \tag{7.7}$$

The actual digital signature consists of the numbers r and s.

After Bob receives the values of r and s, he must perform the following series of calculations:

$$u_1 = s^{-1}h(M) \;(\bmod\; q), \tag{7.8}$$

$$u_2 = s^{-1}r \;(\bmod\; q), \tag{7.9}$$

$$v = g^{u_1}y^{u_2} \bmod p \;(\bmod\; q) \tag{7.10}$$

and verify whether $v = r$.

The fact that $v = r$ must hold is a direct consequence of the following statement:

$$g^{u_1} \cdot y^{u_2} = g^{u_1} \cdot g^{xu_2} = g^{h(M)/s} \cdot g^{rx/s} = g^{(h(M)+rx)/s} = g^k.$$

If $v = r$ then Bob can conclude that the message is still intact. After all, assuming M had changed, then $h(M)$ would also have changed and the value for s which Bob received would be different from the one Alice had calculated with eq. (7.7). Consequently, Bob would compute the wrong values for u_1 and u_2 and discover that v is not equal to r. Whether or not the message originated from Alice follows from the fact that Alice is the only person who knows the value of x (remember that it is practically impossible to derive x from the values of g and y). However, if nobody else knows x, then nobody else is able to generate the correct value for s.

From eqs. (7.8) and (7.9) it follows that it must hold that $s \neq 0 \;(\bmod\; q)$ for verification of the signature to be possible. If the signature is generated and the value of s turns out to be zero, then a new signature must be generated with a different value for k. In practice, however, this will not occur very frequently, since the probability of such a result is of the order of 2^{-160}, taking into account the value of q.

We will demonstrate the algorithm in the following example. To keep the example simple, we will disregard the requirements for the magnitude of the prime numbers.

Example

Select $q = 5$ and $p = 2q + 1 = 11$. With $h = 7$ the value g follows from:

$$g = h^{(p-1)/q} = 7^2 = 5 \ (\mathrm{mod}\ 11).$$

By setting $x = 3$, we find that

$$y = g^x \ (\mathrm{mod}\ p) = 5^3 = 4 \ (\mathrm{mod}\ 11).$$

Assume that Alice chooses $k = 2$. Then it follows that

$$r = g^k \bmod p \ (\mathrm{mod}\ q) = 5^2 \bmod 11 \ (\mathrm{mod}\ 5) = 4.$$

Let $h(M) = 7$, then

$$s = k^{-1} \ (h(M) + xr) \ (\mathrm{mod}\ q) = (7 + 12)/2 \ (\mathrm{mod}\ 5) = 2.$$

Bob's verification will result in

$$u_1 = s^{-1} h(M)(\mathrm{mod}\ q) = 7/2 \ (\mathrm{mod}\ 5) = 1,$$

$$u_2 = s^{-1} r \ (\mathrm{mod}\ q) = 4/2 \ (\mathrm{mod}\ 5) = 2,$$

$$v = g^{u_1} y^{u_2} \bmod p \ (\mathrm{mod}\ q) = 5^1 \cdot 4^2 \bmod 11 \ (\mathrm{mod}\ 5) = 4. \qquad \triangle$$

In general, the DSA is a faster algorithm for generating signatures than the RSA. For verification, however, the opposite is true. Verification of a digital signature generated with the DSA will require more time than verification of a digital signature generated with the RSA. This may prove to be a disadvantage for the DSA, but, in practice, a digital signature is often added to a message only once, whereas verification is performed many times, possibly over a period of years. Obviously, in such a case, fast verification is desirable.

7.6 Zero-knowledge techniques

A development concerned with authentication, especially entity authentication, is the introduction of so-called *zero-knowledge techniques*. Consider the situation in which Alice wishes to prove to Bob that he really is dealing with Alice. Alice could show Bob some kind of identification (for instance a pass-word), so that Bob could verify her identity. The only problem with this is that Bob then knows Alice's identification and can illegally pretend to be Alice. This also applies to a potential intruder who has managed to obtain Alice's identification. Zero-knowledge techniques allow Alice to prove her identity to Bob, without forcing her to reveal any detailed personal information.

In reality, the techniques used are based on a form of entity authentication. They are called zero-knowledge techniques, because after completion of the protocol, Bob will know no more about Alice's means of identification than before.

The first publication on the practical application of zero-knowledge techniques was Fiat and Shamir (1986). It describes a protocol for the identification of one entity by another and a procedure for generating digital signatures.

We will relate our discussion of the basic principle of zero-knowledge to the following example.

Suppose a cave as shown in Figure 7.20 is blocked by a rock. Alice claims that she knows a secret passage to circumvent this obstacle and wishes to convince Bob of this fact, without revealing to him how she actually accomplishes this. No other person besides Alice makes this claim. Bob and Alice make the following agreements. Alice will enter the cave and Bob will follow after a brief delay. If Alice does indeed know of a secret passage, then she must be able to convince Bob that he will not encounter her in the cave.

Alice walks into the cave, while Bob remains at the entrance. Unfortunately, the darkness of the entrance prevents Bob from seeing whether Alice turns right or left. She then proceeds through the passages until she arrives at the obstacle. In the mean time Bob enters the cave and tosses a coin to decide whether to turn right or left. Suppose he turns left. After a while, Alice hears Bob coming and slips through the secret passage to get to the other side of the obstacle. Thus, Bob will not meet Alice in the cave. If Alice had been waiting in the right-hand section of the cave, she would not have used the secret passage. Bob would have remained on the other side of the obstacle and would never be able to find Alice.

If Alice did not know a secret passage, then the probability of running into Bob would be 50%. Therefore Bob will not be convinced immediately by a single attempt as described above. However, if the protocol is repeated several times, Bob will slowly start to believe that Alice really does know of a secret passage, despite never actually seeing it himself. After k attempts, Bob will be convinced and the probability of the existence Alice's secret passage is $1 - 1/2^k$.

The protocol Fiat and Shamir described in their 1986 paper is based on a similar idea. A system is organised around a central body/entity, which provides each member of a certain group (for instance the users of a computer network) with secret personal information. This secret information is generated from a large number n, which is the product of two large prime numbers, p and q, as in the RSA system. The value of n is made public, but the two numbers p and q are known only to the centrally located third party. This central entity will generate an identification sequence I for each member of the group, which contains all the relevant information such as name, address, etc. The central entity can calculate k different integers $v_j = f(I, c_j)$, where c_j is an integer and v_j has an integer value for which

$$u_j^2 = v_j \,(\text{mod } n), \tag{7.11}$$

with u_j between 0 and $n - 1$.

The function $f(.,.)$, which is, in fact, made public, can be realised with, for

Figure 7.20. Alice, Bob and the cave.

example, the triple-DES. The k values can be obtained by repeatedly calculating different values of $f(I,c_j)$, until k values for v_j have been found, which satisfy the above condition. The central entity will then calculate the smallest roots of $\bar{v}_j^{-1}$ (mod n), for each of the k values of v_j. These are denoted by s_j:

$$s_j^2 \, v_j = 1 \ (\text{mod } n). \tag{7.12}$$

Here, we should note that the calculation of these roots requires information about the factors p and q (we will return to this at the end of this chapter). Since we may assume that the factorisation of a large number is impossible (remember our discussion of the RSA system), no one other than the central entity can calculate the values of s_j. Thus, these values are used as the secret values with which others can ascertain a person's identity. The actual values of s_j will provide no information with respect to p and q. Therefore, there is nothing to prevent n being shared by more than one member. After this initialisation phase, the zero-knowledge protocol can commence.

Let us suppose that Bob wishes to ascertain the identity of Alice. Alice must therefore prove in some manner that she has access to the secret values $s_1, ..., s_k$, without actually revealing these values. The protocol requires the following.

Zero-knowledge protocol for identification

(1) Alice sends I and the values $c_1, ..., c_k$ to Bob.
(2) Bob generates $v_j = f(I,c_j)$ for $j = 1, ..., k$.
(3) Alice selects a random number r_i between 0 and $n - 1$ and sends

$$x_i = r_i^2 \ (\text{mod } n) \tag{7.13}$$

to Bob.

(4) Bob generates a binary random vector $(t_{i1},...,t_{ik})$ and sends this to Alice.
(5) Alice sends y_i to Bob, for which

$$y_i = r_i \prod_j s_j^{t_{ij}} \ (\text{mod } n). \tag{7.14}$$

(6) Bob computes

$$z_i = y_i^2 \prod_j v_j^{t_{ij}} \ (\text{mod } n), \tag{7.15}$$

and checks if $z_i = x_i$.

(7) Steps (3)–(6) are repeated for $i = 1, \ldots, t$.

Bob will accept that the person claiming to be Alice is really Alice if all t checks are successful.   ●

Before answering the question of how the given algorithm inequivocally proves Alice's identity, we will first demonstrate that if all proceeds according to plan, then for all i, $z_i = x_i$.

By combining eqs. (7.13)–(7.15), it follows that:

$$z_i = r_i^2 \prod_j s_j^{2t_{ij}} \prod_j v_j^{t_{ij}} \pmod{n}$$

$$= r_i^2 \cdot \prod_j \{s_j^2 \cdot v_j\}^{t_{ij}} \pmod{n}$$

$$= r_i^2 \pmod{n} = x_i. \tag{7.16}$$

Suppose Charles wishes to pretend he is Alice. Since he knows neither the values of s_j, nor the values of r_i, he cannot calculate the values y_i in step (5). Although Charles knows the values of v_j and n, this alone is still not sufficient for calculating s_j. The reason for this is that the square of a value can be calculated without any problem for modulo calculations, but the calculation of a root is impossible if n is sufficiently large. This also applies to the computation of r_i from x_i. Furthermore, Charles will also find it impossible to deduce the values of y_i from eq. (7.15): Charles knows the values of the z_is, since these must be equal to those of x_i, and the values of v_j and the elements t_{ij} of the random vector, but he will still be defeated by the problem of finding a square root of a modulo value.

The following justifies labelling this protocol a zero-knowledge protocol. Consider the most simple case, in which $k = t = 1$. Alice has one secret key s and must demonstrate this to Bob. According to protocol, she generates a random number r, calculates x and sends this value to Bob. Bob will return exactly one bit: a 0 or a 1. If he returns a zero, then Alice will respond by sending r. If he returns a one, then Alice will send the product rs. Bob can verify that Alice has responded correctly by using his knowledge of the values of I and x. If Alice returns r, Bob will not learn anything about s, since r is a random number. And if Alice returns the product rs, he still will not learn anything about s, because now the value of r is not known.

As we reexamine the protocol, we see that instead of generating a random vector in step (4) of the protocol, it is also possible to generate a random matrix T, consisting of the elements t_{ij}, with $i = 1, \ldots, t$ and $j = 1, \ldots, k$. Assuming that at step (3), Alice generates t random values r_i and sends these to Bob, Bob returns the matrix T and Alice generates all t values for y_i, Bob can check the values of y_i as in step (6), in a single round. With the introduction of a random matrix T, which replaces a series of random vectors, the required number of data exchanges is reduced by a factor t.

Before proceeding with an example of the protocol, we will first turn our attention to the calculation of s_i from the expression $s_j^2 v_j = 1 \pmod{n}$, as this appears in the initialisation phase. We must find the square root modulo n. In practice, this is only possible when the prime factors p and q are known. If we know p *and* q, then we can employ the following theorem.

Theorem 7.1

For a given integer value a and with $n = pq$, it holds that the value of w which satisfies

$$a = w^2 \pmod{pq} \tag{7.17}$$

can be derived from the expression

$$w = cu + dy, \tag{7.18}$$

where x and y are solutions to the equations

$$a = x^2 \pmod{p} = y^2 \pmod{q}. \tag{7.19}$$

Furthermore, c and d are given by

$$c = bq \text{ and } d = ap, \tag{7.20}$$

in which a and b are integers for which

$$1 = ap + bq. \tag{7.21}$$

Proof

From eqs. (7.20) and (7.21) it follows that $c = bq = 1 - ap$. Therefore,

$$c = 0 \pmod{q} = 1 \pmod{p}.$$

In the same manner it follows from $d = ap = 1 - bq$ that:

$$d = 0 \pmod{p} = 1 \pmod{q}.$$

Employing $w = cx + dy$ we find that:

$$w^2 = (cx + dy)^2 = c^2x^2 + d^2y^2 + 2cx\,dy = y^2 \pmod{q},$$

but also

$$w^2 = (cx + dy)^2 = c^2x^2 + d^2y^2 + 2cx\,dy = x^2 \;(\text{mod } p).$$

Since $a = x^2 \;(\text{mod } p) = y^2 \;(\text{mod } q)$, it must hold that

$$w^2 = a \;(\text{mod } p) = a \;(\text{mod } q).$$

Thus, we have demonstrated that indeed, w satisfies

$$a = w^2 \;(\text{mod } pq). \qquad \qquad \square$$

This theorem enables us to reduce the problem of finding the roots modulo n to the less complicated problem of finding the root modulo a prime number p or q. If the values of x and y are not immediately evident, special algorithms for finding the root modulo a prime number can be used. Unfortunately, such an algorithm is not available for every prime number. Different algorithms have to be used, depending on the characteristics of the given prime number.

In Section 6.5 we presented the solutions in the case $p = 3 \;(\text{mod } 4)$.

Peralta (1986) provides a probabilistic method for determining x for the case in which p can be written as $p = 2^s r+1$ (r odd, $s > 0$). This procedure consists of three steps.

(i)   Choose an integer $h \leq p - 1$. If $h^2 = a \;(\text{mod } p)$, then $x = h$.
(ii) Compute $(h + \sqrt{a})^{(p-1)/2} = f + g\,\sqrt{a}$.
(iii) If $f = 0$, then $x = g^{-1} \;(\text{mod } p)$ else repeat step i.

It can be explained as follows that this algorithm must produce the sought value of x. If $f = 0$, then step (ii) yields:

$$(h + \sqrt{a})^{(p-1)/2} = g\,\sqrt{a}.$$

Therefore,

$$g^2 a = (h + \sqrt{a})^{(p-1)} = 1 \;(\text{mod } p).$$

The last step is a direct result of Euler's theorem combined with the fact that $h + \sqrt{a}$ is relatively prime with respect to p.

It can be demonstrated that there is a 50% probability of finding a value of h which immediately leads to the solution for x.

We will now illustrate the entire protocol with an example.

Example

Let $n = 33$, $t = 3$ and $k = 2$. The central entity will generate two numbers, v_1 and v_2, based on the identification sequence I. The values of these are 31 and 35, respectively. Note that u_1 and u_2 do indeed exist, so eq. (7.11) is satisfied. Their values are $u_1 = 25$ and $u_2 = 16$.

Now, the values of Alice's secret numbers s_1 and s_2 are computed using v_1 and v_2, which satisfy eq. (7.12). With $v_1 = 31$, we must find a value for s_1, for which $31s_1^2 = 1$ (mod 33). The Euclidean algorithm of Section 6.2 enables us to rewrite this as

$$s_1^2 = 16 \text{ (mod 33)}.$$

Referring to Theorem 7.1, we will first consider

$$x^2 = 16 \text{ (mod 3) and } y^2 = 16 \text{ (mod 11)},$$

in which $p = 3$ and $q = 11$ are chosen.

It is immediately evident that $x = 1$ and $y = 7$ form a possible solution. (If not, then the above algorithm can be employed for finding a root modulo a prime number.)

Since $1 = ap + bq = -7p + 2q$, it follows that $c = bq = 22$ and $d = ap = -21$. Theorem 7.1 leads to $s_1 = w = cx + dy = 22 \times 1 - 21 \times 7 = 7$. In the same manner, we find for s_2 from $25s_2^2 = 1$ (mod 33) that $s_2 = 13$. Thus, Alice's two secret number are (7,13).

We can now proceed with the actual protocol. Alice transmits I, c_1 and c_2 to Bob. Bob uses these values to generate $v_1 = 31$ and $v_2 = 25$, as described in step (2) of the protocol.

Alice chooses the following set of random numbers (12,27,8) and as stipulated by step (3) calculates:

$$x_1 = 12^2 = 12 \text{ (mod 33)},$$

$$x_2 = 27^2 = 3 \text{ (mod 33)},$$

$$x_3 = 8^2 = 31 \text{ (mod 33)}.$$

These values are then sent to Bob. Bob constructs a matrix T:

$$T = \begin{vmatrix} 1 & 0 \\ 0 & 1 \\ 1 & 1 \end{vmatrix}.$$

Alice then uses eq. (7.8) to calculate the values of y_1, y_2 and y_3 as in step (5) of the protocol:

$$y_1 = 12 \times 7^1 \times 13^0 = 18 \ (\text{mod } 33),$$

$$y_2 = 27 \times 7^0 \times 13^1 = 21 \ (\text{mod } 33),$$

$$y_3 = 8 \times 7^1 \times 13^1 = 2 \ (\text{mod } 33),$$

which she subsequently sends to Bob. Bob finally computes the z_is with the aid of eq. (7.15), thus:

$$z_1 = 18^2 \times 31^1 \times 5^0 = 12 \ (\text{mod } 33),$$

$$z_2 = 21^2 \times 31^0 \times 25^1 = 3 \ (\text{mod } 33),$$

$$z_3 = 2^2 \times 31^1 \times 25^1 = 31 \ (\text{mod } 33).$$

These values are identical to those of the x_i.

No-one else can assume Alice's identity. Suppose an intruder, Charles, wishes to trick Bob into believing he is Alice. He claims that $v_1 = 31$ and $v_2 = 25$ are values which are related to his personal identification sequence. For the sake of convenience, let us assume that Charles generated the same numbers as Alice for r_i, $i = 1, 2, 3$, i.e. (12,27,8). Charles then sends the corresponding values of x_i to Bob, i.e. (12,3,31). Bob responds with the same matrix T and at this point Charles starts to get into difficulty.

According to step (5), he must return y_i, $i = 1, 2, 3$ to Bob. However, this requires knowing that $s_1 = 7$ and $s_2 = 13$. Charles does not know these values, however, and he cannot derive them from $v_1 = 31$ and $v_2 = 25$, because Charles cannot solve $s_j^2 v_j = 1 \ (\text{mod } 33), j = 1, 2$. Somehow, he must send three values y_1, y_2 and y_3 to Bob, so Bob can verify that

$$y_1^2 \times 31^1 \times 25^0 = 12 \ (\text{mod } 33),$$

$$y_2^2 \times 31^0 \times 25^1 = 3 \ (\text{mod } 33),$$

$$y_3^2 \times 31^1 \times 25^0 = 31 \ (\text{mod } 33).$$

These equations also do not enable Charles to calculate the values of y_1, y_2 and y_3. Again, we have stumbled upon the problem of finding the root modulo a number. Thus it is impossible for an intruder to assume Alice's identity.

There is another situation which we might consider. After communicating with Alice, Bob may try to pose as Alice to another party, say David. Bob can pretend he is Alice by sending (12,3,31) to David. David will return a matrix T, which may differ from the previous matrix T. Suppose David returns:

$$T = \begin{vmatrix} 0 & 0 \\ 1 & 1 \\ 1 & 0 \end{vmatrix}.$$

If Bob sends the values of y_i, (18,21,2), to David and David checks these, he will find:

$$z_1 = 18^2 \times 31^0 \times 25^0 = 27 \neq 12,$$

etc.

David now knows that somewhere, someone is trying to deceive him. We have seen that it is impossible to find the correct values of y_i which could hide any form of duplicity from Bob's verification. The only situation in which it is possible that David will unjustly conclude that he is communicating with Alice, when in reality he is dealing with Bob, occurs when by chance, David generates the same matrix T as was used previously. If Bob were to send (18,21,2) to David, then all would proceed according to plan. The probability of this event occurring is directly related to the number of possibilities for T. Here, it is equal to 1/64. In general, the probability is equal to $1/2^{kt}$. The larger the matrix, the smaller the probability of accidentally selecting the same matrix.

We can conclude from this discussion that it is indeed possible to ascertain Alice's identity, without disclosing her personal data, the values (s_1, s_2).

In order to keep the probability of mistaking Bob for Alice less than 10^{-6}, the product kt must be greater than 20 ($2^{-20} \approx 10^{-6}$). For instance, $k = 1$ and $t = 20$, or, respectively, 2 and 10, or 4 and 5, etc. Here, the choice of values depends on several factors, such as storage capacity, processing complexity (number of operations) and the complexity of the protocol itself.

An interesting property of the protocol is the fact that it is possible to alter the level of security even after the key (the numbers k and s_j) has been determined. An example is provided by a magnetic identification card containing, for instance, $k = 10$ secret values s_j. If only a fast verification is required, then a single check can be performed ($t = 1$), with a corresponding security level of 2^{-10}. In situations which require a higher level of security, more checks can be made. With $t = 5$, a security level of 2^{-50} is obtained. However, this increase in security does not require any modification of the key on the card itself.

This protocol can be expanded to a protocol for digital signatures. A modification must then be made to Bob's role. In the identification protocol Bob will select a random vector/matrix. When generating digital signatures with zero-knowledge techniques, the unpredictable vector is obtained by applying a Hash function to a message M: $h(M)$.

The following steps must be followed in order to supply a message with a digital signature.

Zero-knowledge protocol for digital signatures

(1) Alice selects random values for r_i between 0 and $n - 1$ ($i = 1, ..., t$) and calculates:

$$x_i = r_i^2 \pmod n, i = 1, ..., t. \tag{7.22}$$

(2) Alice then calculates the Hash result $h(M,x_1,...,x_t)$ and uses the first kt values as the elements t_{ij} of the random matrix T.

(3) For each $i = 1, ..., t$, Alice calculates:

$$y_i = r_i \prod_j s_j^{t_{ij}} \pmod n. \tag{7.23}$$

(4) Alice sends I, M, T and all the y_is to Bob.

Alice's signature for M is verified according to the following steps:

(1) Bob calculates $v_j = f(I,c_j)$ for $j = 1, ..., k$.

(2) Bob then computes for $i = 1, ..., t$:

$$z_i = y_i^2 \prod_j v_j^{t_{ij}} \pmod n. \tag{7.24}$$

(3) Bob finally verifies that the first kt bits of $h(M,z_1,...,z_t)$ are identical to the elements t_{ij}.   ●

Again, it holds that for all values of i: $z_i = x_i$, provided that the protocol was performed correctly. Consequently, it must hold that $h(M,z_1,...,z_t) = h(M,x_1,...,x_t)$, in which the first kt elements correspond to the kt elements of T. The probability of a false message M' producing a Hash result with exactly the same kt elements t_{ij} is given by 2^{-kt}. Here, in practice, we must also base our choice of k and t on the performance desired.

In general, it is beneficial to use zero-knowledge techniques for digital signatures, rather than RSA based methods, especially in those applications where the available processing power is limited. The required number of mathematical operations is far less than for the RSA.

8

Key management and network security

8.1 General aspects of key management

In the previous chapters we discussed cryptographic methods which can be employed for the protection of data. These methods can be utilised to ensure the confidentiality of a message as well as to guarantee the integrity and authenticity (reliability). We have now examined these aspects of generating secure data. One aspect which still remains to be discussed is the importance of the keys employed by these algorithms and methods. Often, the key must be kept secret, since the security of many cryptographic algorithms and methods depends to a large extent on the secrecy of the key, no matter how ingenious and safe the algorithm may be. Whoever has access to the key, can also access the information, assume someone else's identity, etc. This applies not only to symmetrical systems, which require unconditional secrecy of all keys, but also to asymmetrical systems, which are based on both public and secret keys.

These problems are considered in *key management*. In general, we can state that key management is concerned with the management of keys from the point at which they are created to the point at which they are destroyed. Thus, this topic covers:
- generation,
- distribution,
- storage,
- replacement/exchange,
- usage,
- and final destruction

of keys.

A system must not only prevent intruders from obtaining a key, but in addition, it must avoid unauthorised use of keys, deliberate modification and other forms of manipulation of keys, etc. In addition, it is often desirable to be able to detect any situation in which this occurs. Naturally, once the reliability of a key is impaired, its use must be terminated immediately.

This chapter will focus primarily on one aspect of key management, which is concerned with the distribution of keys. Before going into this in more detail, we will first briefly examine the other aspects of key management, mentioned above.

Key generation

First of all, the generated key must be as unpredictable as possible. If certain keys were to occur with a higher probability than others, in practice, a potential intruder would find it easier to discover the used key. This unpredictability of the keys can be ensured by using a pseudorandom generator, such as the DES, which generates a 64-bit pattern.

In practice (for example in the financial world), often a number of people each generate their own random patterns (for instance by tossing a coin), and these together form the key. Let $Z1$ denote a random number generated by a first person and $Z2$ denote a second random number generated by someone else. A key can be found by considering the modulo 2 addition of these two numbers:

$$K = Z1 + Z2. \tag{8.1}$$

If both persons enter their numbers into a security module, which isolates the actual modulo 2 addition from both parties, then neither of the parties will know the secret key K. This principle, according to which two or more persons/systems must collaborate in order to protect vulnerable information and where no party has direct access to the entire information, is often referred to as *dual control.*

Obviously, there are situations in which the generated key must be tested for weakness or semi-weakness, such as is the case for DES. For instance, for the RSA algorithm, the numbers p and q must be tested for primeness and also for their 'strength' as prime numbers.

Key distribution

An obvious method of distribution of keys is simply by hand. This method was frequently used in the days of couriers. Nowadays though, it is used only sporadically, since most key distribution is performed automatically.

Automatic distribution is not only more convenient, but often even essential. For instance, think of a car telephone, which requires two parties to transmit their security keys along the same communication line. The same applies to computer networks, terminal–host connections, etc. In these cases often two types of keys are employed: keys which are used for the actual security of the data and for authentication, etc., the so-called *session keys*, and keys which are used for the security of these session keys during transmission, the so-called *transportation keys*.

Storage of keys

The same comments apply to the storage of keys, as to the transportation of keys. In other words, session keys must be stored in an enciphered form, which requires a *storage key*, comparable to the transportation key employed for key distribution. Transportation keys and storage keys are also referred to as *meta keys*.

Key exchange

It is evident that a regular change of key and in particular the session key is essential for the security of the system. A cryptanalyst can be discouraged in the search for the key, by frequently changing the key. This will also frustrate an exhaustive key search, since by the time the key search succeeds, a different key will already be in use.

Obviously, the frequency with which the key is replaced depends on:

- The sensitivity of the enciphered data: the more important it is to guarantee a high level of security, the more frequently a key will be exchanged for a new one.
- The period of validity of the protected data: for instance, the validity of messages which are communicated via a police telephone is limited; at the time of transmission they can contain valuable information for an intruder. However, a few moments later they will have already lost their news-content.
- The strength of the cryptographic algorithm: the key of an 'ordinary' application of the DES must be exchanged more frequently than the key of a triple-DES system.

Key destruction

This may be performed by simply deleting a key from memory. The deletion of information may be either passive, by removing the supply voltage of the memory chip, or active, by writing over the memory locations.

8.2 Key distribution for asymmetrical systems

When considering key distribution, we must distinguish between asymmetrical and symmetrical algorithms. In this section, an example of key distribution for asymmetrical algorithms is presented, which is based on the RSA algorithm, is presented.

Consider two parties, A and B, whose public keys are P_A and P_B, respectively, and secret keys are denoted by S_A and S_B. Suppose that A wishes to transfer a key K to B. A may do this by first enciphering the key K with his own secret key and subsequently enciphering this result with B's public key. The total result can be written as $(K^{S_A})^{P_B}$. Then B can find the original key K by decrypting the received message with his own secret key followed by a second decipherment, with A's public key. This procedure corresponds with that of Section 7.5, where a method was discussed for enciphering a message including a digital signature (see Figure 7.16).

Here, though, the problem of the authenticity of the public keys still remains unsolved. A enciphers the message with B's public key P_B, regardle of the fact that anyone can assume the identity of B and send A an incorrect public key.

This problem may be solved by introducing a so-called *trusted authority*, for instance a key distribution centre, which issues authentic keys P_A and P_B. When A wishes to communicate with B, he first asks the key distribution centre for B's public key. The key distribution centre key will add a digital signature to this key, so A can use the public key of the key distribution centre for verifying that he has indeed received B's public key.

A different method for exchanging keys in an asymmetrical system is the Diffie–Hellmann protocol. Consider a network of parties/systems who communicate bilaterally in an enciphered form, controlled by a key. This kind of communication will require a large number of keys and managing these keys correctly can be very complicated. The total number of keys is $n(n-1)$ in a system of n persons/subsystems, in which two parties are involved during each communication. Each party must have $n-1$ keys available to be able to communicate with the other $n-1$ parties/systems.

Diffie and Hellmann have devised a method with which the number of keys can be reduced drastically. Their system is based on the exponential and logarithmic functions. Select a prime number p and a primitive element a; both these numbers may be public. Each user A_i may now choose a secret key x_i (smaller than p) and calculate the value of y_i, according to:

$$y_i = a^{x_i} \pmod{p}. \tag{8.2}$$

The values of y_i are made public. Suppose a user A_i wishes to communicate with a user A_j. This will require a mutual key K_{ij}, with which the messages can be enciphered and deciphered. The value of K_{ij} can be computed as follows. User A_i must first calculate

$$K_{ij} = y_j^{x_i} \pmod{p}. \tag{8.3}$$

A_i can perform this calculation, since p is published, x_i is A_i's own key and y_j was disclosed to A_i by A_j. User A_j can calculate the key which is given by the expression

$$K_{ij} = y_i^{x_j} \pmod{p}. \tag{8.4}$$

This will result in exactly the same key for both parties, due to the symmetry of the power function. We find that

$$K_{ij} = y_j^{x_i} \pmod{p} = (a^{x_j})^{x_i} \pmod{p}$$

$$= (a^{x_i})^{x_j} \pmod{p} = y_i^{x_j} \pmod{p}. \tag{8.5}$$

Anyone trying to tap into this communication will have considerably more difficulty in finding the key K_{ij}. Relying only on the known values of p, y_i and y_j, he must try to solve:

$$K_{ij} = y_i^{\log y_j} \bmod p = y_j^{\log y_i} \pmod{p}, \tag{8.6}$$

in which the base of the logarithm is a. The security of the system is guaranteed by the fact that the intruder is unable to compute $\log y \pmod{p}$. Thus, we are again relying on a one-way function, whose value can easily be calculated, but whose inverse value is impossible to compute, as was the case for the RSA algorithm.

The strength of this procedure lies in the fact that it no longer requires n^2 different keys, but only n keys; one key for each user. A disadvantage of this system though, is that authenticity is no longer ensured. Suppose an intruder A^* manages to intercept the value of y_j, which A_j sends to A_i. A^* has selected a secret value for x^* and therefore, his public number is given by $y^* = a^{x^*} \pmod{p}$. If A^* were to send this value of y^* to A_j, then A_j would believe that this is the value of y_i, sent by A_i. He would then continue to communicate with A^*, under the false assumption that he is communicating with A_j. In order to protect oneself against this kind of malpractice, some form of authentication of the public values of y_i must be introduced.

8.3 Key distribution for symmetrical algorithms

Key distribution for symmetrical algorithms will initially always involve physically distributing a set of keys.

For instance, the installation of a network will require a security officer to load (initial) keys into so-called *tamper-resistant modules*, which provide physical protection for the keys stored in these modules. After this initial distribution, the actual distribution, i.e. the distribution of the session keys, can be carried out over the network. The key distribution centre can fulfil two functions. First of all, for instance during installation, the key distribution centre can prescribe which users can communicate with each other and which cannot, by supplying the users with the keys of only those with whom they may communicate. This is referred to as an *off-line key distribution system*.

In an *on-line key distribution system*, each user has an individual and unique key for communicating with the key distribution centre. Before communication between two parties using the system can take place, the key distribution centre must be requested for the relevant session/communication key. In this case, each user will need to store only a small number of keys, whereas for an off-line key distribution system, the number of keys stored per user is proportional to the total number of users. The following is an example of a protocol for on-line key distribution.

(1) A informs the key distribution centre of the wish to communicate with B, by means of encryption with a secret key $K1$, which was distributed previously. Thus, A sends the following message to the key distribution centre: $A//EK1[A//B]$.

(2) The key distribution centre can decipher this message, as it has access to all of the keys distributed. It can therefore deduce that A wishes to communicate with B and composes the following message: $EK1[KS//EK2[KS]]$, in which KS represents the secret key for the communication between A and B (the session key).

(3) A now sends $EK2[KS]$, as received from the key distribution centre, to B along with his own address. B concludes that the message was transmitted by A and can retrieve the secret session key KS with the aid of the secret key $K2$, which he shares with the key distribution centre. A and B can now finally communicate with the secret key KS, known only to themselves.

This system is in fact a *hierarchical key system*: it involves keys $(K1,K2)$ for transporting another key KS, which is used for the actual communication.

A good example of a *hierarchical key management system*, based on enciphered keys, is the IBM key management scheme. Consider a system of a large number of terminals connected to a central computer (host). The data which are transported between the host and terminals must be enciphered with keys which are stored in the host. These keys must be transported along the same communication lines as the data. We will now explain exactly how the IBM key management scheme deals with this problem.

The IBM key management scheme relies on four different keys: the master keys $KM0$ and $KM1$, the terminal key KT and the session key KS. The key KS is used for enciphering the actual data which is communicated between host and terminal and is continuously changed. The key KT is used for transporting the key KS from the host to the terminal. The keys $KM0$ and $KM1$ are known only within the host and are used for ensuring that neither KS nor KT appears outside the security modules in an unprotected form.

Figure 8.1 demonstrates how the data to/from the terminal is enciphered/deciphered within the host. The message $EKM0(KS)$ is to be interpreted as the encipherment of the key KS, with the master key $KM0$. This is deciphered in a tamper-resistant module, thus allowing data received from the terminal to be decrypted and data destined for the terminal to be enciphered. As mentioned previously, the key KS is sent to the terminal in an encrypted form. This step is performed within the host in a module illustrated by Figure 8.2.

Figure 8.1. (*a*) Host tamper-resistant encipher module, (*b*) host tamper-resistant decipher module.

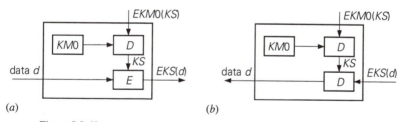

(*a*)   (*b*)

Figure 8.2. Host tamper-resistant module for generating the enciphered session key.

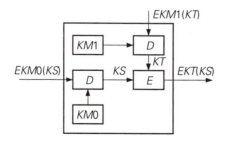

Inside the tamper-resistant module, the keys KS and KT are extracted from $EKM0(KS)$ and $EKM1(KT)$. Then KS is enciphered with KT, resulting in $EKT(KS)$. The keys KT and KS must never appear outside the special crypto-modules in an unprotected form and therefore, the following procedure is employed. In the host a random number R is generated and treated as the ciphertext $R = EKM0(KS)$. The key KS can be derived within the module by deciphering KS with $KM0$ (see Figure 8.2). Therefore, the key KS remains unknown outside the security modules, even within the host! Figure 8.3 illustrates how data is enciphered and deciphered in a terminal, based on $EKT(KS)$.

We may wonder why two master keys are used instead of one. Suppose that KT is also enciphered with the key $KM0$, then it is possible to obtain KS with the tamper-resistant decipher module in the host. See Figure 8.4.

In general, it can be stated that designing key management schemes is not an easy task. As the scheme becomes more complex, the designer will find it very difficult to predict the weak spots in the design in order to arrive at a sufficiently secure system. A trend has developed for employing techniques from the field of artificial intelligence, such as expert systems, for testing new key management systems (see for instance Van der Lubbe and Boekee (1989, 1990)).

We must remark that in practice, solutions implemented in only hardware and software are almost never adequate. Certain keys will always require

Figure 8.3(*a*) Encipherment and (*b*) decipherment in the tamper-resistant modules of the terminal.

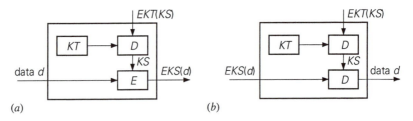

(*a*)   (*b*)

Figure 8.4. One master key instead of two can result in the decipherment of the key KS.

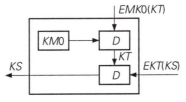

physical protection. That is why the previous example is based on tamper-resistant security modules, which provide physical protection against unauthorised access. In the IBM scheme, the modules protect the master keys $KM0$ and $KM1$ and the terminal key KT from being read illegally. Often, a reinforced cabinet is used which prevents or at least impedes an intruder from gaining access to the contents. In addition, means are provided for detecting any break-in, so that classified information can be deleted from memory before it is disclosed.

Physical obstructions against unauthorised intruders are usually constructed from special materials, such as special metals, certain synthetic resins, etc. However, it must be remembered that there are almost always means of getting around these measures; most metals can be penetrated by lasers and almost all resins can be dissolved by the right chemicals. Therefore, it is essential that any illegal entry can be detected, in order to minimise any possible damage.

Inside the module, sensors may be mounted for detecting an intruder, which can activate an alarm circuit. These sensors may respond to vibration, temperature variations, light, etc.

8.4 Network security

In our previous discussion on key management, we repeatedly came across implementations for computer networks. The keys and data are distributed over this network. Although we will not go into any details on the specific cryptographic aspects of network security, we will briefly present the different types of encipherment which can be implemented in a network. A characteristic of network communication is that data normally travel via several stations, rather than following the most direct route between two parties. Examples of this are computer networks, in which the terminals are connected to each other via a central computer, and a telephone network where the subscribers are connected via an exchange. Often, the description of this kind of network is based on distinguishing between nodes and (communications) links. The nodes, which are interconnected via the links, represent the points at which data is entered, enciphered, stored temporarily or processed in any other way.

If we wish to safeguard the data circulating within the network, we can employ one of three methods:

(1) link encipherment,
(2) node encipherment,

(3) end-to-end encipherment.

In general, the information which is to be transmitted from one node to another consists of the actual message and a header, which provides the necessary routing information, such as the address of the sender, the destination, the format, etc.

Link encipherment requires both the message and the header to be enciphered. Each connection between two nodes is designated its own specific key (see Figure 8.5). As the information travels through the network, it is decrypted at every node in order to read the header and to determine which node it should be sent to next. Then, the information is encrypted with a new key, belonging to the link with the next node and the data are sent on.

The danger of this strategy lies in the fact that at every node, the information is deciphered and then reenciphered with a different key. Thus, for a brief moment, the information is available as plaintext, which can prove a great risk.

Node encipherment is similar to link encipherment. However, for node encipherment, the information encrypted with the first key is not deciphered and then reenciphered at every node, but is transformed directly into a ciphertext based on the new key (see Figure 8.6).

Finally, end-to-end encipherment only encrypts the actual message. The information in the header is added as plain text and the message is not decrypted until it has reached its final destination (see Figure 8.7). A major advantage of end-to-end encipherment compared to the other methods is that

Figure 8.5. Link encipherment.

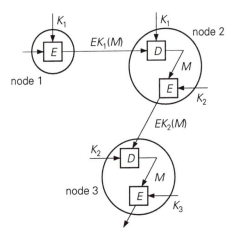

it can provide a much higher level of security. The confidential information will circulate within the network only in an enciphered form and only the transmitter and the receiver have access to the key. This implies that a user must have access to the keys of every possible destination, or at least must be able to request these from a key distribution centre. On the other hand, link and node encipherment only require knowledge of the keys which are necessary for communicating with neighbouring nodes.

Another advantage of end-to-end encipherment is that it uses far fewer cryptographic operations, as here, the inter-lying nodes are only used as relay stations.

A disadvantage of end-to-end encipherment compared to the other methods is the availability of the header information as plain text throughout the entire network. This means a third party will be capable of so-called *traffic flow analysis*, i.e. he can monitor which users are exchanging

Figure 8.6. Node encipherment.

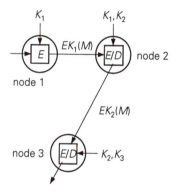

Figure 8.7. End-to-end encipherment.

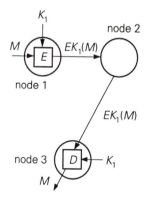

information and when. This can be frustrated by occasionally sending dummy messages over the network.

8.5 Fair cryptosystems

It is obvious that cryptography provides an effective way of achieving privacy and security in communication. Unfortunately though, the reverse side of the coin is that organised crime is also increasingly using cryptography. One way to stop this trend would be to ban all use of cryptography in public networks. However, this would immediately cause numerous practical problems, since many sectors of our society are already entirely dependent on cryptography; for instance automatic banking, communication within multi-national companies, etc. However, even apart from that, a ban on cryptography is in direct conflict with a person's desire and right to privacy. Also, it is clear that organised crime would not take much notice of such a ban on the use of cryptography. It is important, though, to get more control of organised crime, since indirectly, it is still a threat to any democratic society and therefore to the privacy of each individual citizen.

One can consider the use of so-called *weak cryptosystems*. These have a built-in 'trapdoor', which makes it possible for a public authority, for instance, to tap a communication channel. Also, systems can be considered, whose keys can be found by an *exhaustive key search*, provided the investment necessary for performing such a search is only within the reach of a national government. A third possibility is to force anyone who wishes to use cryptography to supply the keys to a specially authorised body. However, with all these methods, the person's privacy will still be at risk.

Thus, we are looking for cryptosystems which can guarantee a person's right to privacy on the one hand and on the other also provide a means for a public authority to tap the system, if necessary. This kind of system is often referred to as a *fair cryptosystem* (Micali 1993). These systems are capable of ensuring the rightful user's privacy, but also of disregarding that of an illegal user (i.e. a criminal). Fair cryptosystems allow an authorised body (which has been appointed by the government after consulting the people) to obtain the plaintext of the enciphered data, even without the permission and/or knowledge of the user. This is only allowed under special circumstances, which have been defined by, for instance, a court of law.

The basic idea is that the user divides his key into say, five parts and deposits each of these five parts with the appointed body (the trusted party). These five parts must be such that (*a*) the original key can be reconstructed from these five parts, (*b*) if only four or less parts are available, it is impossible to reconstruct the key and (*c*) it is possible to determine for each part separately, whether or not it is correct. We will use the Diffie–Hellmann protocol as an example and adapt it, so that it can be used as a fair cryptosystem.

Fair Diffie–Helllmann cryptosystem

Let p be a prime number and a some kind of primitive element. User A_i selects five random integers $x_{i1}, ..., x_{i5}$, for which $x_{ih} \leq p - 1$, for all values of h. The secret key belonging to A_i, denoted by x_i, is equal to the sum modulo p of the five integers. The public key y_i is

$$y_i = a_i{}^x \pmod{p},$$

just as in the original protocol. User A_i must also calculate:

$$t_1 = a^{x_{i1}}, ..., t_5 = a^{x_{i5}}$$

It can easily be seen that

$$t_1 t_2 ... t_5 = a^{(x_{i1} + ... + x_{i5})} = a^{x_i} = y_i.$$

Let $T_1, T_2, ..., T_5$ represent the five trusted parties. Then A_i must give: t_1 and x_{i1} to T_1; t_2 and x_{i2} to T_2, etc. The parts $t_1, ..., t_5$ can be made public, although x_{i1} must be known only to T_1, etc.

Upon receiving t_1 and x_{i1}, T_1 must verify that indeed $a^{x_{i1}} = t_1$. If this is the case, then he will store (y_i, x_{i1}). Finally, he will add a digital signature to (y_i, t_1) and pass this signed pair on to a key management centre.

The key management centre then verifies that the product of $t_1, ..., t_5$ does actually produce the public key y_i, which is then also signed digitally and made available to the user, as was the case for the original protocol.

The user A_j sets to work in a similar fashion. A_j also generates five integers and determines the secret key x_j. The five trusted parties are supplied with the relevant information and at the end of the process the key management centre will release the public key y_i.

The rest of the protocol proceeds in the same manner as the original protocol. The public and secret keys are used to determine K_{ij}, which is used as the actual key for encrypting the communication between A_i and A_j.

The validity of this protocol can be explained as follows. Just as in the original protocol, if an intruder only knows the public key, it will still be impossible to obtain the secret keys and therefore he will not be able to find

K_{ij}. The security of this protocol relies on the fact that it is impossible to determine a logarithm modulo a number.

The trusted party T_1 has a copy of x_{i1}, $t_1 = a^{x_{i1}}$ and y_i. However, this is not sufficient to calculate the secret key x_i, since in order to be able to do that, he would also have to own a copy of x_{i2}, ..., x_{i5}. Even if four of the parties decided to co-operate, together they would not be able to calculate the secret key x_i. The key management centre also does not have sufficient information to reconstruct the secret key. Although it knows t_1, ..., t_5 and y_i, it still cannot calculate x_i. Even if it were to conspire with four of the trusted parties, still no result could be obtained. Only if, for instance, a court order forced all trusted parties to release x_{i1}, ..., x_{i5}, would it then be possible to derive x_i and thus K_{ij}.

If a criminal were to use the protocol as explained above and supply information to the trusted parties in the correct manner, then he runs the risk that the encrypted information will be deciphered if he ever were to become suspected of illegal activities. Furthermore, if he were to use a different cryptosystem or to supply incorrect information to the trusted parties, then he has to expect to answer to the authorities for using an illegal form of enciphering, when his messages proved impossible to be deciphered.

Micali (1993) explains how the RSA system can also be adapted for use as a fair crypto-system.

We will conclude this section with some remarks with respect to an American project, started in 1993, under the name Skip-jack, for the development of a fair crypto-system (referred to here as a *key-escrow system*) for the American telephone network. Each telephone is equipped with a chip containing a secret symmetric cryptographic algorithm: the Skip-jack algorithm. The actual algorithm is secret, although it works with 64-bit blocks and is assumed to resemble the DES, only with a higher level of security. In addition to the algorithm, the chip also contains a unique identification number I which is 30 bits long, and a unique key K_I which is 80 bits long. The secret key K_I is reconstructed from two parts, each also 80 bits long. Each part is released to a trusted party (often referred to as the *escrow agent* in this context), together with the identification number of the chip. Suppose that two parties wish to communicate. Then they will use a session key K_S. The chip will not only transmit the encrypted data (enciphered with K_S), but also the session key, after it has been enciphered with the secret key of the chip, $E_{K_I}(K_S)$, and the chip's identification number. This additional information comprises the so-called law enforcement access field (LEAF). A law enforcement agency can tap the communication line and use the LEAF to derive the chip identification number I. The agency can then request each trusted party to hand over their part of the key K_I, provided that there are sound legal grounds for such a request. Once the two parts of the key are available, the agency can reconstruct the key K_I and thus decipher the encrypted key K_S. It is obvious, however, that once K_I has been deciphered, it is possible to decipher any following K_S. Therefore, the original concept also includes provisions which allow tapping only for a limited period of time.

Appendix A

Shannon's information measure

In 1948 Claude E. Shannon laid the foundation for what is currently referred to as information or communication theory, with the publication of his article 'A mathematical theory of communication'. A year later he presented his article 'Communication theory of secrecy systems' (Shannon 1949), which is an important contribution to cryptography.

Information theory is the science concerned with the concept of information, how to measure information and its applications. It hinges on the information measure, as defined by Shannon, which is related to uncertainty by the notion of probability. Before defining Shannon's information measure, we must introduce several methods of notation.

Consider an information source X, which generates a series of symbols. We will assume that the symbols are selected from a source alphabet, given as

$$X = (x_1, x_2, \ldots, x_n).$$

For each symbol, a probability of occurrence is defined. The probability corresponding to a symbol x_i is denoted as $p(x_i)$. The entire set of probabilities of the elements of X, the so-called probability distribution, is denoted by

$$P = \{p(x_1), p(x_2), \ldots, p(x_n)\}.$$

In the same manner, we can define a second information source Y, with its related alphabet, as

$$Y = (y_1, y_2, \ldots, y_m),$$

and the corresponding probability distribution as

$$Q = \{q(y_1), q(y_2), \ldots, q(y_m)\}.$$

207

The probability of the information source X producing a symbol $x_i \in X$, while the information source Y produces a source $y_j \in Y$, is written as $p(x_i, y_j)$. The conditional probability of x_i given the occurrence of y_j is defined as

$$p(x_i/y_j) = p(x_i, y_j)/\ q(y_j) \text{ provided } q(y_j) > 0.$$

This represents the probability that information source X will generate a symbol x_i, when information source Y has generated a symbol y_j. Hence, it assumes there is some form of dependency between the two sources.

Correspondingly, we can define the conditional probability of y_j given the occurrence of x_i as

$$q(y_j/x_i) = p(x_i, y_j)/\ p(x_i) \text{ provided } p(x_i) > 0.$$

The amount of information or uncertainty with respect to the information source X, the *marginal information measure*, is defined by Shannon as

$$H(X) = -\sum_{i=1}^{n} p(x_i) \log p(x_i). \tag{A.1}$$

For an alphabet consisting of only two symbols, with $p(x_1) = p$ and $p(x_2) = 1 - p$, this formula will lead to the following expression for the measure of information:

$$H(X) = -p \log p - (1 - p) \log (1 - p).$$

In Figure A.1 the behaviour of $H(X)$ as a function of p is plotted. We can

Figure A.1. $H(X)$ for the case of two symbols.

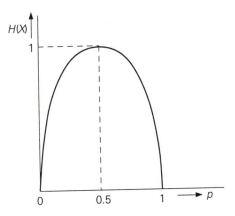

conclude from this figure that when the uncertainty of a symbol is zero, i.e. the probability of occurrence is 1, then the information measure becomes zero. This corresponds to the intuitive notion that an assured event does not provide any new information or uncertainty. The same also applies when $p = 0$; in this case the other of the two symbols will occur with a probability 1. When $p = 0.5$, $H(X)$ will reach its maximum value, which is equal to 1 bit, if the base of the logarithm is set to 2. In this case, the probability of both symbols is the same and we have complete uncertainty as far as the occurrence of a symbol is concerned. Thus, this occurrence will provide maximum information. (Note by definition, $0 \log(0) = 0$.)

For the general case with an alphabet of n symbols, the information measure reaches a maximum when the probability of occurrence is equal for all symbols and reaches a minimum when the probability of one of the symbols is 1. It holds that

$$0 \leq H(X) \leq \log n, \tag{A.2}$$

where the lower and upper limits correspond to the minimum and maximum values for the information measure. In the same manner, we can define an information measure or uncertainty $H(Y)$ for a given source Y. This is left as an exercise for the reader.

The definition of the information measure related to a pair of symbols (x_i, y_j), the *joint information*, corresponds to that of the marginal information measure:

$$H(X,Y) = -\sum_{i=1}^{n}\sum_{j=1}^{m} p(x_i, y_j) \log p(x_i, y_j). \tag{A.3}$$

This formula holds for the general case, in which there may be some form of dependence between the two information sources. However, if the two sources are independent, i.e. the occurrence of a symbol of one source is in no way influenced by the occurrence of symbols of the other source, we find that:

for all i and j: $p(x_i, y_j) = p(x_i).q(y_j)$,

and thus

$$H(X,Y) = H(X) + H(Y). \tag{A.4}$$

Finally, we may also define a *conditional information measure* $H(X/Y)$, which is a measure of the uncertainty of the symbols of source X, when the symbols of Y are given.

Analogous to the definition of the marginal information measure, the information measure with respect to X given a symbol y_j can be defined as

$$H(X/y_j) = -\sum_{i=1}^{n} p(x_i/y_j) \log p(x_i/y_j). \tag{A.5}$$

By taking the average for all values of y, we arrive at the mean information measure of X, with prescience of the symbols generated by Y,

$$H(X/Y) = -\sum_{i=1}^{n}\sum_{j=1}^{m} p(x_i,y_j) \log p(x_i,y_j). \tag{A.6}$$

$H(Y/X)$ may be defined in the same manner. It can be demonstrated that

$$H(X/Y) \le H(X). \tag{A.7}$$

This implies that with foreknowledge of the symbols of source Y, the uncertainty in the symbols of source X is less than or equal to the uncertainty in the case in which we do not have prior knowledge of the symbols of Y.

The equality applies when the two sources are totally independent. In this case, information about one source does not supply any information on the other source. The quantity

$$I(X;Y) = H(X) - H(X/Y) \tag{A.8}$$

is called the *mutual information measure*. Referring to the previous inequality, we see that this expression may be used as a measure of the (in)dependence between X and Y. There is a direct relation between the marginal, the conditional and the combined information and thus also the mutual information. This is given by

$$H(X,Y) = H(X) + H(Y/X)$$

$$= H(Y) + H(X/Y). \tag{A.9}$$

The above formula expresses the fact that the combined information is the sum of the marginal information and the conditional information. The expression we found earlier, which relates the marginal and the conditional information measures, enables us to write

$$H(X,Y) = H(X) + H(Y/X) \le H(X) + H(Y), \tag{A.10}$$

in which the equation holds when X and Y are independent. Thus, we can say that the combined measure of information reaches its maximum when both information sources are independent and decreases as the dependency increases. In the case of complete dependency, the symbols of Y are known as soon as those of X are given, and therefore $H(Y/X) = 0$. This leads to

$$H(X,Y) = H(X). \tag{A.11}$$

The definitions and expression as given above form the basis of the field of information theory. With the aid of these formulae, we can gain insight into various aspects of information transportation and information storage. In addition, they enable us to devise the most efficient methods of transportation and storage. Finally, Chapter 3 of this book clearly demonstrates the importance of information theory in the field of cryptology.

Appendix B
Encipherment of imagery

In the previous chapters we have assumed that the messages which are to be enciphered are always one-dimensional. However, this is not necessarily the case for many kinds of data transport and storage. Image processing, for instance, often uses two-dimensional or even three-dimensional signals. This occurs, for example, in a videophone, which produces a sequence of images. Here, the third dimension is time. A second example is provided by weather satellites, which contain equipment for generating various images of the earth's atmosphere from different bands of the spectrum. Here, the third dimension is the frequency of radiation.

This type of multi-dimensional signal is generally characterised by a high level of redundancy. When considering images, for instance, we see that they often contain large areas in which the pixels (picture elements) have more or less the same grey value/colour etc. In particular, a videophone image will normally represent the face of the talking person in front of a monochromatic background and the third dimension especially will contribute to the high level of redundancy. The differences between consecutive images of a videophone will be small; only parts of the face and possibly the shoulders and arms may move, producing (local) changes to the image.

Due to the redundancy in the images, if we were to use uncompressed images for encipherment, the algorithm would be required to handle vast amounts of data. Moreover, certain algorithms are unsuitable, such as for instance, DES in the ECB mode, since this algorithm always produces the same ciphertexts with a given input sequence of 64 bits. Therefore, an image of a plain background would result in a series of identical blocks of data, which a potential intruder may use to acquire information on the original image. If the image data is compressed and reduced before encipherment, obviously the redundancy of the enciphered images can be reduced considerably.

Another problem when enciphering multi-dimensional data is the speed desired when processing the algorithms. Uncompressed images are usually order of 2^{19}–2^{25} bits and for a sequence of the images, the required speed of encipherment amounts to somewhere between ten and a few hundred Mbits/s. Even when data reduction and compression techniques are applied, the remaining volume of data still requires the use of high speed cryptographic algorithms.

Therefore, in addition to work on more general cryptographic algorithms, such as the DES, algorithms specifically designed for enciphering images also currently form an important area of research.

Scrambling

A popular method of concealing the information in an image is by scrambling. This involves altering the format of the signal, often also removing the field and line pulses of an image and enciphering and transmitting these separately. An example of this for TV signals is provided by Davidov, Bhaskaran and Wechselberger (1984). The television signal is divided into three parts: an analogue scrambled picture, control data and a digital audio signal. The latter two signals are encrypted by means of the DES in the CFB mode.

Random field

Kafri and Keren (1987) have constructed a method of encipherment especially for binary images: i.e. the pixels in the image are either black or white, represented by the values 0 or 1, and cannot assume any intermediate grey values. Suppose the image is $m \times n$ pixels and that, in addition, we have a totally random field also of $m \times n$ pixels. Later, we will explain exactly how this field is generated. The random field is used as the key, with which the original image is enciphered, according to the following procedure.

– If the value of the pixel of the original image M is 1, then the value of the

Figure B.1. Scrambling of television signals.

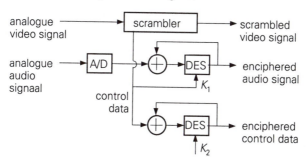

associated pixel of the cipher image C is equal to the value of the corresponding pixel in the random field K.

– If the value of a pixel in M is 0, then the value of the pixel in the cipher image C is equal to the value of the complement of the corresponding pixel of K.

Thus, we can express this with the logical formula: $C = \neg\,(M \vee K)$.

Consider the following example of a section of an image M and the corresponding section of the random field K.

$$M:\ \begin{array}{|cccc|} \hline 1 & 0 & 1 & 0 \\ 0 & 1 & 0 & 1 \\ 1 & 0 & 1 & 0 \\ 0 & 1 & 0 & 1 \\ \hline \end{array} \qquad K:\ \begin{array}{|cccc|} \hline 0 & 0 & 1 & 0 \\ 1 & 0 & 0 & 1 \\ 1 & 1 & 0 & 1 \\ 0 & 1 & 0 & 1 \\ \hline \end{array}$$

The locations in M which contain a 1 produce the corresponding value of K. For the locations of M containing a 0, we must calculate the complement of the corresponding value of K. This will result in the cipher image C.

$$C:\ \begin{array}{|cccc|} \hline 0 & & 1 & \\ & 0 & & 1 \\ 1 & & 0 & \\ & 1 & & 1 \\ \hline \end{array} + \begin{array}{|cccc|} \hline & 1 & & 1 \\ 0 & & 1 & \\ & 0 & & 0 \\ 1 & & 1 & \\ \hline \end{array} = \begin{array}{|cccc|} \hline 0 & 1 & 1 & 1 \\ 0 & 0 & 1 & 1 \\ 1 & 0 & 0 & 0 \\ 1 & 1 & 1 & 1 \\ \hline \end{array}$$

The image can be deciphered by reversing the procedure: at those locations of K containing a 1, the corresponding value of C remains unchanged. At the locations of K containing a 0, the value of the deciphered pixel is equal to the complement of the corresponding pixel of C. Therefore,

$$M:\ \begin{array}{|cccc|} \hline & & 1 & \\ 0 & & 1 & \\ 1 & 0 & & 0 \\ & 1 & & 1 \\ \hline \end{array} + \begin{array}{|cccc|} \hline 1 & 0 & & 0 \\ & 1 & 0 & \\ & & 1 & \\ 0 & & 0 & \\ \hline \end{array} = \begin{array}{|cccc|} \hline 1 & 0 & 1 & 0 \\ 0 & 1 & 0 & 1 \\ 1 & 0 & 1 & 0 \\ 0 & 1 & 0 & 1 \\ \hline \end{array}$$

This method is in fact a two-dimensional implementation of stream encipherment. Therefore, the random fields can be generated with the aid of shift registers, which construct the random field line by line.

This method may also be adapted for enciphering grey-value images. The grey value of each pixel is represented by a binary value. Thus in the case of an $m \times n$ image with 256 grey values, the dimension of the image is expanded to $8m \times n$ pixels. Obviously, the random field must also be expanded to corresponding dimensions.

Area-filling curves

Matias and Shamir (1987) have developed a method for enciphering images, based on so-called area-filling curves. The image is scanned according to a given curve, thus rearranging the order of the pixels. This method is therefore basically a transposition with a period equal to the size of the image. The resulting image represents the cipher image and the curve along which the image is scanned forms the key.

Methods of constructing an area-filling curve, i.e. a curve which passes through all the pixels of the image once, are described in this section.

The image is regarded as a grid or raster, in which each point represents a picture element. The grid is divided into closed segments by joining the points of the grid with horizontal and vertical lines, see Figure B.2. Finally, all the segments are opened and connected by two parallel lines, resulting in a single curve filling the entire area and passing through each of the pixels of the area.

An alternative method of generating an area-filling curve is by constructing it line by line. See Figure B.3. Here, the curve is composed by connecting the grid points with horizontal and vertical lines, working through the grid line by line. Each extension of the curve must be such that the ends of the resulting fragments can always be continued onto the next line of the grid. When the bottom line of the grid is reached, the ends of all

Figure B.2 Generating area-filling curves: (*a*) the grid (*b*) segments (*c*) area-filling curve.

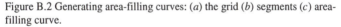

(*a*)   (*b*)   (*c*)

Figure B.3. Generation of area-filling curves: resulting fragments down to line i, $i + 1$ and $i + 2$.

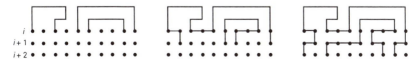

the fragments are connected and the curve is closed.

Another method for generating area-filling curves is described by the following. This method relies on so-called basic elements, which represent a curve in a block of $k \times k$ grid points. The curve connects all the grid points in this area and has one starting point and one end point. The beginning and the end of the curve must be positioned in order to allow for the beginning and end of two neighbouring elements to be joined. See Figure B.4.

Let us consider a basic element of 3×3 pixels, in order to gain an idea of the order of magnitude of this principle. A basic element of 3×3 can contain two possible curves (see Figure B.5(a)). In a 6×6 grid, it is possible to construct $2^4 = 16$ curves based on these two basic elements (see Figure B.5(b)). For an $m \times n$ image, the total number of different curves is $2^{mn/9}$. Thus, for example, an image of 256×256 pixels, the total number of possibilities is in the order of 10^{2000}.

It is evident that the number of possible curves and, consequently, the total number of keys is extremely large. The weakness of area-filling curves does not lie in the number of possible curves, but in the number of neighbouring pixels. Suppose we are considering pixel (i,j), then we can always conclude that the preceding pixel of the curve can be one of only four possibilities, i.e. $(i,j-1)$, $(i,j+1)$, $(i-1,j)$ or $(i+1,j)$, regardless of the shape or size of the area-filling curve. Furthermore, this pixel will always be succeeded by one of the remaining three pixels. Therefore, if the method of area-filling curves is employed for enciphering a sequence of images, in which the differences between consecutive images are only small, as is the case for videophones, then a given order of pixels, say from (i,j) via $(i,j-1)$ to $(i+1,j-1)$, will occur repeatedly. Bertilsson, Brickell and Ingemarsson (1989) have demonstrated that this weakness can be used by cryptanalysts. Combinations of neighbouring pixels which do not occur frequently are related to edges in the image. Bertilsson *et al.* managed to locate these in a set of 25 consecutive

Figure B.4. Generation of area-filling curves: (a) basic element, (b) area-filling curve constructed from a number of basic elements.

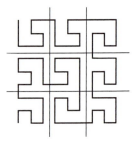

(a)   (b)

images and thus reconstruct parts of the contours in the image. Normally, almost all the information of the image is contained in the contours.

The procedure may be improved by increasing the number of neighbouring candidate pixels, for instance by allowing for diagonal displacements in addition to horizontal and vertical displacements. The principle of the algorithm remains unchanged, however. Therefore, a cryptanalyst will only require a few more images of a sequence to obtain the same result. This also applies when the procedure is extended from two-dimensional area-filling curves to three-dimensional curves, where the third dimension is time. The number of neighbouring pixels will increase to 26, but this still does not exclude a cryptanalytic attack as described above. Consider, for instance, a series of videophone images. If 30 images are generated per second and the image remains more or less the same for 5 seconds, then we will have 150 almost identical images. The number of times pixel (i,j) is followed by pixel $(i,j-1)$ will on average be $150/26 = 6$. This is sufficient to allow us to locate (parts of) the contours within the image.

Figure B.5. Generation of area-filling curves: (a) two basic elements of a 3×3 grid, (b) all resulting curves within a 6×6 grid.

(a)   (b)

The only real solution to this problem is to generate pseudorandom curves. When tracking a pseudorandom curve, the distance and direction of a successor pixel will remain as random as possible.

A binary pseudorandom sequence may be generated with the aid of a shift register. If a sequence is divided into k-bit words, where k depends on the dimensions of the image, then a pair of consecutive k-bit words may be regarded as a representation of the x and y co-ordinates of the pixels. The words of the sequence will point to a new pixel as the pseudorandom sequence is processed. By filling a new image line by line with the indicated pixels, a new enciphered image is obtained (see Figure B.6).

It can be demonstrated that the distance and direction of the displacements between two consecutive pixels of the enciphered image correspond to those of the ideal (random) case. (See also Van der Lubbe, Spaanderman and Boekee (1990a).)

Figure B.6. Image transposition based on pseudorandom sequences: (*a*) original, (*b*) enciphered image.

(*a*)   (*b*)

Bibliography

Abramowitz, M., Stegun, I.A.: *Handbook of Mathematical Functions*; with formulas, graphs and mathematical tables, Dover Publications, New York, 1965.

Andronicos, M., *et al.: The Greek Museums*, Ekdotike Athenon, Athene, 1975.

Beker, H.J., Piper, F.C., Communications security; a survey of cryptography, *IEEE Proc.*, **129**, Pt A no 6, 1982, 357-76.

Bertilsson, M., Brickell, E.F., Ingemarsson, I.: Cryptanalysis of video encryption based on space-filling curves, in *Proc. EUROCRYPT '89, Houthalen, Belgium, 10-13 April 1989*, Springer-Verlag, Berlin, 1989.

Beth, T.H., Hess, P., Wirl, K.: Kryptographie; *Eine Einführung in die Methoden und Verfahren der geheimen Nachrichtenübermittlung*, Teubner, Stuttgart, 1983.

Biham, E., Shamir, A.: Differential cryptanalysis of DES-like cryptosystems, Internal Report, the Weizmann Institute of Science, Dept. Applied Mathematics, March 25, 1990.

Biham, E., Shamir, A.: Differential cryptanalysis of the full 16-round DES, Technical Report 708, Technion – Israel Institute of Technology, Department of Computer Science, December 1991.

Blom, R.: On pure ciphers, Internal Report, Li-TH-I-0286, Linkoping University, Sweden, 1979.

Boekee, D.E., Van Tilburg, J.: The Pe-security distance as a measure of cryptographic performance, in *Proc. Int. Symp. Information Processing and Management of Uncertainty in Knowledge-Based Systems, Paris, June 30-July 4, 1986*, pp. 130-133.

Boekee, D.E., Van der Lubbe, J.C.A.: Error probabilities and transposition ciphers, in *Proc. Ninth Symposium on Information Theory in the Benelux, Mierlo, the Netherlands, May 26-27, 1988*, Werkgemeenschap voor Informatie- en Communicatietheorie, Enschede, K.A. Schouhamer Immink (ed.), 1988, pp. 155–62.

Brickell, E.F., Odlyzko, A.M.: Cryptanalysis: a survey of recent results,: *IEEE Proc.* **76**, no 5, May 1988, 578–93.

Davidov, M., Bhaskaran, V., Wechselberger, T.: Commercial applications of encrypted signals. In: *MILCOM '84, IEEE Military Communications Conf.*, Vol. 2, October 1984, pp. 307–12.

Davies, D.W., Price, W.L.: *Security for Computer Networks; An introduction to data security in teleprocessing and electronic funds transfer*, Wiley, Chichester, New York, 1989.

Davio, M., *et al.*: Analytical characteristics of the DES, Internal Report, Philips Research Laboratory, Brussels, Belgium, October 1983.

Diffie, W.: The first ten years of public-key cryptography, *IEEE Proc.*, **76**, no 5, May 1988, 560–77.

Diffie, W., Hellman, M.E.: Exhaustive cryptanalysis of the NBS Data Encryption Standard. In: Computer, June 1977a, 74–84.

Diffie, W., Hellman, M.E.: New directions in cryptography, *IEEE Trans. Inform. Theory*, **IT-22**, no 6, November 1977b, 644–54.

Fiat, A., Shamir, A.: How to prove yourself: Practical solutions to identification and signature problems, in *Advances in Cryptology – Crypto '86*, Lecture Notes in Computer Science 263, A.M. Odlyzko (ed.), Springer-Verlag, Berlin, 1986, pp. 186–94.

Feige, U., Fiat, A., Shamir, A.: Zero-knowledge proofs of identity, *J. Cryptology*, **1**, 1988, 79–94.

Feigenbaum, J.: Overview of Interactive Proof Systems and Zero-Knowledge, in *Contemporary Cryptology: The Science of Information Integrity*, G.J. Simmons (ed.), IEEE Press, New York, 1991, pp. 423–39.

Fumy, W., Riess, H.P.: Kryptographie; *Entwurf und Analyse symmetrischer Kryptosysteme*, Oldenbourg Verlag, Munchen, Wien, 1988.

Goldwasser, S., Micali, S., Rackoff, C.: The knowledge complexity of interactive proof systems, *SIAM J. Comput.*, **18**, no 1, February 1989, 186–208.

Golomb, S.W.: *Shift Register Sequences*, Holden-Day, San Francisco, 1967.

Golomb, S.W.: *Shift Register Sequences*, Aegean Park Press, 1984.

Gordon, J.: Strong RSA Keys, *Electronic Letters*, 7th June 1984, **20**, no 12, 514–16.

Grafisch Museum Drenthe: Het raadsel van Phaistos, brochure ter gelegenheid van de Nationale Wetenschapsdag, 16 oktober 1988.

Gutekunst, T.: Der Lucifer-Algorithmus, *Mikro- und Kleincomputer*, **8**, no 4, 1986, 55–60.

Hardy, G.H., Littlewood, J.E., Polya, G.: *Inequalities*, Cambridge University Press, 1973.

Hellman, M.E.: An overview of public key cryptography, *IEEE Commun. Society Magazine*, November 1978, 24–31.

Jansen, C.J.A.: Investigations on nonlinear streamcipher systems: construction and evaluation methods, Ph. D. Thesis, Delft University of Technology, Delft, 1989.

Kafri, O., Keren, E.: Encryption of pictures and shapes by random grids, *Optic Letters*, **12**, 1987, 377–9.

Koblitz, N.: Elliptic curve cryptsystems, *Math. of Computation*, **48**, 1987, pp. 203–209.

Koblitz, N.: *A course in number theory and cryptography,* Springer-Verlag, New York, 1988.

Kranakis, E.: *Primality and Cryptography*, J. Wiley and Sons, Chicester, New York, 1986.

Lai, X.: Detailed description of a software implementation of the IPES cipher, unpublished, 1991.

Leiss, E.L.: *Principles of data security*, Plenum Press, New York, 1982.

Lenstra, A.K., Manasse, M.S.: Factoring by electronic mail, in *Advances in Cryptology – Proc. Eurocrypt '89,* Lecture Notes in Computer Science 434, J.J. Quisquater and J. Vandewalle (eds.), Springer-Verlag, Berlin, 1990, pp. 355–71.

Lenstra, A.K., Manasse, M.S.: Factoring with two large primes, in *Advances in Cryptology–Proc. Eurocrypt '90,* Springer-Verlag, Berlin, 1991, pp. 72–82.

Lenstra, H.W.: Integer programming with a fixed number of variables,: *Math. Operations Res.,* **8**, no 4, November 1983, 538--48.

Lenstra, H.W.: Primality testing, in *Computional Methods in Number Theory,* H.W. Lenstra and R. Tijdeman (eds.), Mathematical Centre Tracts, 154, Vol. 1, 1982, Mathematisch Centrum Amsterdam.

Lenstra, H.W.: Primality Testing Algorithms, *Séminaire Bourbaki,* **33**, no 576, 1980/81, 243–57, SVLNM, number 901.

Martin, J.: *Security, accuracy, and privacy in computer systems,* Prentice-Hall, Englewood Cliffs, New Yersey, 1973.

Massey, J.L.: Shift-register synthesis and BCH decoding, *IEEE Trans. in Inform. Theory,* **IT-15**, no. 1, January 1969, 122–7.

Massey, J.L.: An introduction to contemporary cryptology. In: *IEEE Proc.* **76**, no 5, May 1988, 533–49.

Massey, J.L.: The relevance of information theory to modern cryptography, in *Proc. Bilkent Int. Conf. on New trends in Communication, Control and Signal Processing (BILCON'90), Ankara, Turkey, July 2-5, 1990,* Elsevier Science Publ., Amsterdam, 1990, pp. 176–82.

Matias, Y., Shamir, A.: A video scrambling technique based on space filling curves, in *Advances in Cryptology – Crypto '87,* Lecture Notes in Computer Science, Springer-Verlag, Athens, 1987, pp. 398–417.

Maurer, U.M.: A provably-secure strongly-randomized cipher, presented at Monte Verita Seminar on Future Directions in Cryptography, Ascona, Switserland, October 15–21, 1989.

Menezes, A.J.: *Elliptic Curve Public Key Cryptosystems,* Kluwer, Boston, 1993

Merkle, R., Hellman, M.: Hiding Information and Signatures in Trapoor Knapsacks, *IEEE Trans. Inform. Theory*, **IT-24-5**, September 1978.

Merkle, R.C., Hellman, M.E.: On the security of multiple encryption, *Comm. of the ACM*, **24**, 7, 465, July 1981, 465–7.

Meyer, C.H., Matyas, S.M.: *Cryptography: A New Dimension in Computer Data Security*, John Wiley & Sons, New York, 1982.

Micali, S.: Fair cryptosystems, Report MIT/LCS/TR-579b, Laboratory for Computer Science, MIT, 1993.

Miller, V.: Uses of elliptic curves in cryptography, in *Advances in Cryptography – Crypto '85,* Lecture Notes in Computer Science, **218**, Springer-Verlag, Berlin, 1986, pp. 17–26.

Peralta, R.C.: A simple and fast probabilistic algorithm for computing square roots modulo a prime number, *IEEE Trans. Inform. Theory*, **IT-32**, no 6, November 1986, 846–7.

Pfleeger, Ch. P.: *Security in Computing*, Prentice-Hall, Englewood Cliffs, New Yersey, 1989.

Pless, V.S.: Mathematical foundations of interconnected J-K flip-flops, *Inform. Control*, **30**, 1976, 128–42.

Poe, E.A.: De gouden kever, in: *Fantastische Vertellingen*, Utrecht, 1973, pp. 7–38.

Pope, M.: *The story of decipherment; from Egyptian hieroglyphic to linear B*, Thames and Hudson, London, 1975.

Riesel, H.: *Prime Numbers and Computer Methods for Factorization*, Birkhauser, Boston, 1985.

Rivest, R.L.: RSA chips (past/present/future). In: *Advances in Cryptology – Crypto '84*, Lecture Notes in Computer Science 209, T. Beth and I. Ingemarsson (eds.), Springer-Verlag, Berlin, 1985, pp. 159–65.

Rivest, R., Shamir, A., Adleman, L.: A Method for Obtaining digital Signatures and Public-Key Cryptosystems, in *Comm. of the ACM*. **21**, 1978, 120–6.

Rueppel, R.: *Analysis and design of stream ciphers*, Springer-Verlag, Berlin, 1986.

Shamir, A.: A polynomial-time algorithm for breaking the basic Merkle-Hellman cryptosystem, *IEEE Trans. Inform. Theory*, **IT-30**, no 5, September 1984, 699–704.

Shannon, C.E.: The mathematical theory of communication, *Bell Syst. Tech. J.*, **27**, 1948, 379–423, 623–56.

Shannon, C.E.: Communications theory of secrecy systems, *Bell Syst. Tech. J.*, **28**, 1949, 656–715.

Siegenthaler, T.: Decrypting a class of stream ciphers using ciphertext only, *IEEE Trans. Computers*, **C-34**, 1985, 81–5.

Simmons, G.J. (ed): *Contemporary Cryptology: The Science of Information Integrity*, IEEE Press, New York, 1991.

Simmons, G.J. (ed.): Special section on cryptology, *IEEE Proc.*, May 1988, 533–627.

Solovay, R., Strassen, V.: A fast Monte Carlo test for primality, *SIAM J. Comp.*, **6**, 1977, 84–5, *erratum* **7**, 1978, 118.

Van der Lubbe, J.C.A., Boekee, D.E.: Een expertsysteem voor key-management-schema's, in *AI Toepassingen '89, Proceedings tweede Nederlandse Conferentie georganiseerd door de werkgroep Expertsystemen*, H.J. van den Herik (ed.), SIC, Amsterdam, 1989, 265–73.

Van der Lubbe, J.C.A., Boekee, D.E.: KEYMEX: An expert system for the design of expert systems, in *Advances in Cryptology, Proceedings Int. Conf. on Cryptology, Sydney, Australia, January 1990*, J. Seberry and J. Pieprzyk (eds.), Lecture Notes in Computer Science 453, Springer-Verlag, Berlin, pp. 96–103.

Van der Lubbe, J.C.A., Spaanderman, J.J., Boekee, D.E.: On cryptosystems for digital imagery, in *Proc. Eleventh Symposium on Information Theory in the Benelux, October 25–26, 1990, Noordwijkerhout, the Netherlands*, J.C.A. van der Lubbe (ed.), Werkgemeenschap voor Informatie- en Communicatietheorie, Enschede, 1990a, pp. 60–6.

Van der Lubbe, J.C.A.: *Information Theory*, Cambridge University Press, Cambridge, 1997.

Van Tilburg, J., Boekee, D.E.: Divergence bounds on key equivocation and error probability in cryptanalysis, in *Advances in Cryptology – Crypto '85*,

Lecture Notes in Computer Science 218, G. Good and J. Hartmanis (eds.), Springer-Verlag, Berlin, 1986, pp. 489–513.

Welsh, D.: *Codes and Cryptography*, Clarendon Press, Oxford, 1988.

Index